Integrated Distributed Intelligent Systems in Manufacturing

Intelligent Manufacturing Series

Series Editor: Andrew Kusiak
Department of Industrial Engineering
The University of Iowa, USA

Manufacturing has been issued a great challenge - the challenge of Artificial Intelligence (AI). We are witnessing the proliferation of applications of AI in industry, ranging from finance and marketing to design and manufacturing processes. AI tools have been incorporated into computer-aided design and shop-floor operations software, as well as entering use in logistics systems.

The success of AI in manufacturing can be measured by its growing number of applications, releases of new software products and in the many conferences and new publications. This series has been established in response to these developments, and will include books on topics such as:

- design for manufacturing
- concurrent engineering
- process planning
- production planning and scheduling
- programming languages and environments
- design, operations and management of intelligent systems

Some of the titles are more theoretical in nature, while others emphasize an industrial perspective. Books dealing with the most recent developments will be edited by leaders in the particular fields. In areas that are more established, books written by recognized authors are planned.

We are confident that the titles in the series will be appreciated by students entering the field of intelligent manufacturing, academics, design and manufacturing managers, system engineers, analysts and programmers.

Titles available

Object-oriented Software for Manufacturing Systems
Edited by S. Adiga

Integrated Distributed Intelligence Systems in Manufacturing
M. Rao, Q. Wang and J. Cha

Artificial Neural Networks for Intelligent Manufacturing
Edited by C. H. Dagli

Integrated Distributed Intelligent Systems in Manufacturing

Ming Rao PhD

Department of Chemical Engineering
The University of Alberta
Canada

Qun Wang PhD

Department of Chemical Engineering
The University of Alberta
Canada

and

Jianzhong Cha PhD

Department of Mechanical Engineering
Tianjin University
Peoples' Republic of China

 CHAPMAN & HALL
London · Glasgow · New York · Tokyo · Melbourne · Madras

Published by Chapman & Hall, 2–6 Boundary Row, London SE1 8HN

Chapman & Hall 2–6 Boundary Row, London SE1 8HN, UK

Blackie Academic & Professional, Wester Cleddens Road, Bishopbriggs, Glasgow G64 2NZ, UK

Chapman & Hall Inc., 29 West 35th Street, New York NY10001, USA

Chapman & Hall Japan, Thomson Publishing Japan, Hirakawacho Nemoto Building, 6F, 1–7–11 Hirakawa-cho, Chiyoda-ku, Tokyo 102, Japan

Chapman & Hall Australia, Thomas Nelson Australia, 102 Dodds Street, South Melbourne, Victoria 3205, Australia

Chapman & Hall India, R. Seshadri, 32 Second Main Road, CIT East, Madras 600 035, India

First edition 1993

© 1993 Ming Rao, Qun Wang and Jianzhong Cha

Printed in Great Britain at the University Press, Cambridge

ISBN 0 412 54370 2

A catalogue record for this book is available from the British Library

Library of Congress Cataloging-in-Publication data available

♾ Printed on permanent acid-free text paper, manufactured in accordance with the proposed ANSI/NISO Z 39.48-199X and ANSI Z 39.48-1984

Contents

Preface

The growing complexity of industrial manufacturing and the need for higher efficiency, greater flexibility, better product quality, and lower cost have changed the face of manufacturing practice. Meanwhile, the development and applications of computer hardware and software techniques have allowed the implementation of advanced manufacturing technology. The core of this modern manufacturing technology is to implement the integration of domains of specific knowledge, the integration of industrial manufacturing, managing and marketing activities as well as decision-making automation. The development and application of integrated distributed intelligent system techniques will quicken this process and improve our production efficiency, product quality, and company competition in sharing the international marketplace. Advanced manufacturing technology is a hybrid generated by the *knowledge* from manufacturing, computers, management, marketing, and systems science. It has an interdisciplinary nature, and allows the application of different knowledge from other scientific fields to manufacturing processes.

In this book, we present the concept, methodology, and implementation techniques of an integrated distributed intelligent system (IDIS), as well as IDIS applications in real-world manufacturing industries. IDIS is a large knowledge integration environment, which consists of several symbolic reasoning systems, numerical computation packages, neural networks, database management systems, computer graphics programs, an intelligent multimedia interface and a meta-system. The integrated software environment allows the running of programs written in different languages, and communication among the programs as well as exchange of data between programs and database. These isolated intelligent systems, numerical packages and models are under the control of a supervising intelligent system, namely the meta-system. The meta-system manages the selection, coordination, operation and communication of these programs. In this book, the structure, implementation and applications of IDIS are discussed.

This book consists of eight chapters. The first part, comprising three chapters, introduces fundamental concepts, methodology and implementation of integrated distributed intelligent systems (IDIS). IDIS applications are described in the following five chapters, that discuss, respectively, an integrated distributed intelligent design environment (IDIDE), an integrated distributed intelligent simulation environment (IDISE), a gear integrated distributed intelligent manufacturing system (GIDIMS), an intelligent system for process startup automation, and an intelligent operation support system for a chemical pulp process (IOSS).

Professor Andrew Kusiak, the editor of the *Intelligent Manufacturing Series*, provided us with very important suggestions and assistance in preparing the book. We would like to acknowledge our friends for their support in the development of this book. Special thanks are due to Dr Ji Zhou (Huazhong University of Science and Technology) for allowing us to use some research results from his laboratory. We greatly appreciate Professor H. Qiu, Dr Y. Shu, Dr H. Yang and Dr D. Li, who helped us by reviewing the manuscript, and providing many useful comments and suggestions for improvement of the book.

Graduate students (J. Corbin, R. Dong, H. Kim, M. Stevenson, Y. Ying, H. Zhou, J. Sun, H. Farzadeh, A. Ursenbach and J. Zhu) and postdoctoral fellows (P. Du, X. Shen, D. Wang, Q. Xia) who work in the Intelligence Engineering Laboratory at the University of Alberta, have made important contributions to the technical content as well as the preparation of this manuscript. We would like to express our appreciation for their help and contributions.

We gratefully acknowledge the financial support from Natural Sciences and Engineering Research Council of Canada and the National Natural Science Foundation of China.

Dr Ming Rao
Dr Qun Wang
Dr Jianzhong Cha

Edmonton, Alberta, Canada

1

Introduction

1.1 Computer-integrated manufacturing system

Computer Integrated Manufacturing is a strategic thrust, an operating philosophy. Its objective is to achieve greater efficiencies within the business, across the whole cycle of product design, manufacturing, and marketing.

—— **F. Greenwood (1989)** ——

1.1.1 International competition and manufacturing industry

International competition has intensified the requirement for high quality products that can compete in the global marketplace. As a result of this increased competition, the pace of product or system development has been quickened, thus forcing manufacturers into an era in which continuous quality improvement is a matter of survival, not simply competitive advantage. As the time scale of the product life cycle has decreased and the demand for quality increased, attention has focused on improving product quality and promoting the competitive ability of companies through better design, manufacturing, management and marketing. Obviously, increasing manufacturing rate and improving product quality will become the most important factors for sharing the international marketplace.

Global manufacturing industry now is undergoing a rapid structural change. As this process continues, our manufacturing industry is encountering difficulties as it confronts a changed and more competitive environment and marketplace. Clearly, we cannot maintain our industrial base and standard of living without an efficient manufacturing industry.

To join the global competition, to share the international marketplace, as well as to increase our total factor productivity (TFP) and manufacturing labor productivity (MLP), our manufactures may implement two strategies. The first is to change the way we do things now, i.e. to improve our enterprise management. The second one is to develop and apply high technology, i.e. computer-integrated manufacturing (CIM) technology and integrated information management (IIM) for our companies and enterprises. Computer-aided design (CAD) defines and describes products on a video screen; computer-aided engineering (CAE) analyzes production performance and productivity; and computer-aided manufacturing (CAM) automates the shop floor process. As a result, faster, cheaper, safer and better production and operation can be achieved.

1.1.2 Development of manufacturing industry

It is widely recognized that the course of industrialization has substantially been the process of automation (Lu, 1989). From a viewpoint of automation, the development of industry automation can be divided into four stages as shown in Table 1.1.

Table 1.1 Development of industrial automation

Stage	Feature	Automation	Design	Manufacture
1	Labor-intensive	None	Individual	Manual
2	Equipment-intensive	Instruments	Group	NC, CNC
3	Information-intensive	Information	CAD	FMS
4	Knowledge-intensive	Decision	ICAD	CIMS

Stage 1. Labor-intensive industry: At this stage, the efficiency and quality of production mainly rely on the skills of human operators using simple machines without automatic controls and operations. Equipment maintenance heavily depends on 'private' experience.

Stage 2. Equipment-intensive industry: Automatic equipment plays a dominant role in the competition of productivity. The equipment may consist of quite complicated mechanical, electronic and computerized devices, and can be automatically used only as a stand-alone machine. A typical example is the numerical control (NC) machine that is highly automatic as a single machine. Meanwhile, sensor techniques and advanced instruments have become important means to obtain industrial data for equipment maintenance and automatic control. As a result of more powerful and affordable computing facilities on factory floors, our industry is now moving into the third stage.

Stage 3. Information-intensive industry: This stimulates the development of flexible manufacturing systems (FMS). At this stage, automation is realized at the level of data processing for groups of automatic machines. CAD/CAM technology is a typical example representing the characteristics of the stage. Based on mathematical models, computers assist human experts for numerical analysis, synthesis, simulation and graphics, and also provide information for them to make decisions. Fault diagnosis and equipment maintenance depend not only on the information from sensors but also on human experts' experience.

With the rapid growth of complexity of manufacturing processes and the demand for higher efficiency, greater flexibility, better product quality, and lower cost, industrial practice has approached the more advanced level of automation. Nowadays, much attention has been given by both industry and academia to computer-integrated manufacturing systems (CIMS). What CIMS try to do is to integrate all stages in a product life cycle, including taking orders from customers, production planning, product design, process scheduling, NC coding, product testing, sales, service and production management, into a complete (or integrated) information processing system with minimum interference to the operation process by human operators.

Stage 4. Knowledge-intensive industry: At this stage, computers help human experts not only for data processing, but also for decision-making. The tendency to replace human brain power by computers is developing into decision-making automation, which is based on the technology of knowledge processes (Lu, 1989). In future industrial companies, intelligent computer-aided design (ICAD) will automatically generate and analyze product models with the help of human expertise; CIMS technology will be used to implement plant-wide integration; and intelligent maintenance systems (IMS) will maintain the computerized automated production processes and equipment. This high-performance automation represents the next generation of manufacturing systems. So far, many efforts have been made to investigate the architecture of the integrated environment for CIMS, since it is recognized as the key issue for this very comprehensive and sophisticated system.

1.1.3 Computer-integrated manufacturing system

Computer-integrated manufacturing (CIM) is a strategic thrust, an operating philosophy. Its objective is to achieve greater efficiencies within the entire business, across the whole cycle of product design, manufacturing and marketing (Greenwood, 1989). Currently, there exist islands of automation, such as computer-aided design, computer-aided manufacturing, flexible manufacturing systems, manufacturing resource planning, office automation, computerized marketed forecasting, and so on. Manual operations and paperwork systems as well as human decision-making power link the islands together. CIM digitally connects these islands by providing fast, accurate, consistent data from decision-making processes. CIM thereby can cut costs, enhance quality, reduce response time, and improve white-collar productivity. As a result, the competitive capacity of a product or company can gain an increased share in the international marketplace.

1.1.4 Problem definition in manufacturing industry

As we know, manufacturing industry has played an important role in the entire industrial development since the industrial revolution in the eighteenth century. However, with the advent of modern industry, especially the applications and development of computer techniques, the use of classical manufacturing techniques makes it difficult to fulfill people's expectations when facing current international competition. On the other hand, in implementing a computer-integrated manufacturing system, many problems may arise:

- Too much information requires manipulation, especially with a newly implemented system. The stochastic occurrence of operational faults requires emergency handling in industrial manufacturing processes. For example, in the chemical industry, this issue is critical because of degraded operator efficiency and quality in handling emergencies, and due to the ever increasing degree of coupling between the system components and the response time requirements. We need to develop the automation of safe process startup and shutdown under irregular conditions to improve production efficiency and safety.

- In many industrial companies, manufacturing processes are modern highly-computerized human–machine systems. Meanwhile, it is becoming increasingly difficult for operators to understand the many signals and information from display board, video, and computer screen. Owing to the importance of the operator's role in these systems, the quality of interaction between human operators and computers is crucial. Thus, development of an intelligent multimedia interface will better facilitate human interaction with complex real-time monitoring and control systems. Through the multimedia interface, computer systems can communicate with operators via multimedia and modes, such as natural language, graphics, animation, and video presentation.

- In a computer-integrated manufacturing system, conflicting conclusions among production sections and different knowledge domains usually arise because different intelligent systems can make different decisions based on different criteria, even though the data that fire the rules are nearly the same. For example, when a disturbance occurs, the *operation expert* may change the operation state in a chemical process, but the *control expert* may wish to keep the operation state unchanged (due to the set-point control strategy).

The lack of conflict-reasoning strategies has become the bottleneck in applications of integrated techniques. In addition, plant-wide optimization requires coordination among different production workshops and sections. Therefore, the integration of management, manufacturing and marketing information is a very important factor.

- The amount of manufacturing information and management information is increasing fast. In manufacturing processes, a large amount of operational data and information from different sensors have to be collected and processed in time with useful knowledge and information acquired. Here, the key problems are (i) how to process so much data in time, and (ii) how to extract *effective knowledge* which is useful for decision-making from new data, because all data cannot be stored in a computer due to excessive amount. Existing database technology and numerical computing methods may be used for operational data processing. However, how to effectively acquire knowledge from the data still remains difficult. Generally speaking, most of the problems involved in industrial manufacturing are usually ill-structured, and difficult to formulate. Some important parameters cannot be measured on-line. Meanwhile, industrial manufacturing very often deals with uncertain and fuzzy information. In such processes, mathematical modeling is not amenable, and purely algorithmic methods are difficult to use.

- In manufacturing processes, there is too much information needing to be represented. Some may be expressed explicitly in graphics or neural networks, rather than numerical or symbolic. So far, a few intelligent systems have been successful in integrating several expert systems and coupling symbolic reasoning with numerical computation (Rao, 1991). However, such integration still cannot satisfy the requirements of CIMS. The empirical data from process operations cannot be effectively processed with the existing expert systems or numerical models. Some process variables that affect process operation, product quality, and product quantity are only partially understood. Sometimes, these variables are not directly measurable. In such a case, neural networks are an alternative to solve the problem. In fact, many complicated industrial problems cannot be solved by a single technique such as symbolic reasoning, numerical computation, or neural networks. A new methodology to integrate neural networks, symbolic reasoning, numerical computation, and computer graphics has to be developed to handle these problems. In addition, modern industrial processes are so complicated that no single tool can handle everything. For example, in an integrated operation environment, different existing software packages coded in computer languages such as C™, FORTRAN™ and Pascal™, as well as written with commercial AI tools such as KEE™, OPS5™, G2™ and M.1™, can be used together.

- Nowadays, most process plants have highly automatic facilities. To stimulate industrial companies to utilize the newest manufacturing technology, we need to develop a system architecture that does not replace the existing factory facilities, operation environments, or information management systems in industrial plants, but effectively integrates and utilizes the knowledge and facilities that are currently available in companies.

To overcome the difficulties and solve the problems above, new techniques and methodologies should be introduced into the real industrial manufacturing environment.

1.2 Artificial intelligence in manufacturing industry

New information techniques will provide the foundation for manufacturing systems of the 1990's. As these systems evolve, both the basic control systems and company operations will change. In response to competitive thrusts increasingly based on flexibility, the integration of manufacturing islands will evolve and necessitate successful applications of knowledge system technology. The effective integration of the AI technology will face challenges that are organizational logistical and technical in nature.

—— **J. Davis and M.D. Oliff (1988)** ——

1.2.1 Background

The growing complexity of industrial manufacturing and the need for higher efficiency, greater flexibility, better product quality, and lower cost have changed the face of manufacturing practice. Meanwhile, applications of computers have allowed the implementation of more advanced manufacturing techniques. Modern manufacturing technology is a hybrid generated by the *knowledge* from manufacturing, computer, management, marketing, and system science as shown in Figure 1.1. Advanced manufacturing technology is interdisciplinary in nature, and allows the applications of different knowledge from other scientific fields to manufacturing processes.

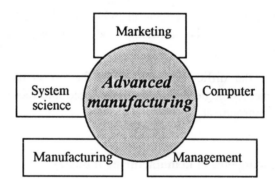

Figure 1.1 Knowledge integration.

For a long time, computers have been used for design, manufacturing, control, diagnosis and monitoring in manufacturing industries. With the advent of computer technology, manufacturing methodologies are developed rapidly. The applications of more powerful computers have allowed us to implement more advanced manufacturing concepts. The systematic knowledge of industrial manufacturing has created an environment facilitating the introduction of artificial intelligence. Existing computer hardware and programming technologies have matured to implement more complicated and flexible manufacturing systems (Bitran and Papageorge, 1987).

On the other hand, the increasing demand for more effective information processing methods and manufacturing strategies in industrial applications cannot be met by current computing techniques.

The technological advances in manufacturing have addressed the research interests of integrated distributed intelligent manufacturing systems, which encompasses the theory and applications of distributed artificial intelligence (DAI), computers and industrial manufacturing technologies (Figure 1.2). Intelligent manufacturing technology is developed for implementing complicated industrial automation, improving product quality, enhancing industrial productivity, ensuring operational safety and protecting the environment. Two objectives of intelligent manufacturing technology are:

- enhancing product quality through automation, and
- improving manufacturing efficiency through artificial intelligence.

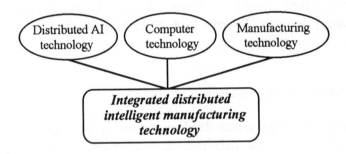

Figure 1.2 Intelligent manufacturing technology.

Currently, computers have been widely used in engineering applications, but the use has been limited almost exclusively to purely algorithmic solution. In fact, many engineering problems are not amenable to purely algorithmic computation. They are usually ill-structured problems that deal with non-numerical or non-algorithmic information and are suitable for the use of AI techniques.

As a new advanced technology frontier, AI has been widely applied to various disciplines, including industrial manufacturing. It aims at processing non-numerical information, using heuristics and simulating a human being's capability in problem-solving. In fact, a much better terminology for this field was suggested as *complex information processing system*, rather than *artificial intelligence*. Today, instead of continuing arguing which terminology is better, we concentrate on investigating the use of powerful computation technology in AI research.

Expert system (or knowledge-based system) technology is one of the most active branches in AI research. An expert system is also a computer program that acquires the knowledge of human experts and applies it to make inferences for users with less training or experience in solving various problems. Expert systems provide a programming methodology for solving ill-structured engineering problems that are difficult to handle by purely algorithmic methods. An expert system is so constructed that it does allow us to capture the way that people reason and think. The experience from developing expert systems for engineering problems has shown that their power is most apparent when the considered problem is sufficiently complex. By means of AI techniques, the search could be reduced to less options.

1.2.2 Intelligence engineering

Traditionally, expert system development mainly relies on the knowledge engineer who, by the definition, is a computer scientist and has a knowledge of artificial intelligence and computer programming. Knowledge engineers interview domain experts to acquire the knowledge, then build expert systems. With more powerful hardware platforms, more user-friendly programming environments, as well as the increased computation capability of our engineers, a new era of AI applications is coming to enable engineers to program expert systems to handle their problems.

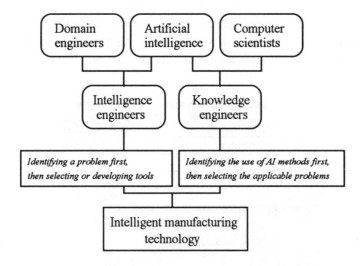

Figure 1.3 Knowledge engineers and intelligence engineers
in intelligent manufacturing.

As we know, during the development of expert systems, knowledge acquisition is the most important, but most difficult task. Even with the help of knowledge engineers, some private knowledge (such as heuristics and personal experience as well as rules of thumb) may still be difficult to transfer. A new generation of engineers, namely intelligence engineers, must be trained to meet the needs in this subject. The name intelligence engineer was set to distinguish them from the knowledge engineer. As shown in Figure 1.3, an intelligence engineer is a domain engineer (for instance, a process engineer), who has certain domain knowledge in the related application area. Through a comparatively short-period training process, she/he learns the basic AI techniques, and gets the hands-on experience of programming expert systems. Thus, the intelligence engineer can build up much better expert systems to solve her/his domain problems, and extend AI applications successfully. In intelligence engineering, a problem to be solved is identified first. Then, the problem-solving methods or tools are selected or developed. AI methodology will not be chosen unless there is a real need. On the other hand, AI methodology has been selected first in the knowledge engineering approach, which results in a very common failure in developing expert systems: the selected tools are not suitable for the problem.

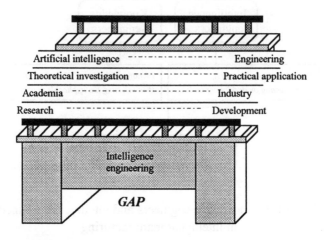

Figure 1.4 Activities of intelligence engineering.

Intelligence engineering involves applying artificial intelligence techniques to solve engineering problems, and investigating theoretical fundamentals of AI techniques and other branches of computer science based on engineering methodology. Figure 1.4 demonstrates the objectives of intelligence engineering. Distinguished from knowledge engineering, intelligence engineering emphasizes integrating knowledge from different disciplines and domains to solve real engineering problems.

In the process of building expert systems, acquisition and representation of knowledge are two of the most important steps. The methodology used in this process is different from that in the prototype problem or in the related quantitative simulation program. In the process of developing an expert system, some new methods to solve the problem may be generated, which can complement the solver of the prototype problem (Rao *et al.*, 1988a and 1988c). As usual, the new technique of programming expert systems is sought to guarantee the realization of new algorithms for this problem. This indicates that building an expert system is not just a translation from the existing expertise knowledge into a computer program; it is the production process in which new expertise knowledge can be acquired. For example, IDSCA was developed to assist designing the controllers in multiloop control systems (Rao *et al.*, 1988a and 1988c). This intelligent system seems to be relatively simple but effective for the design of industrial process control systems. The significance behind the practical application is the investigation of both the new design criterion and an adaptive feedback testing system (AFTS). The former may complement the knowledge of process control, while the later will stimulate the development of intelligent systems with the high reliability and performance.

1.2.3 Intelligent system applications in manufacturing

Applications of intelligent systems in industrial manufacturing can be divided into intelligent design, operation, control, production planning and maintenance (Figure 1.5).

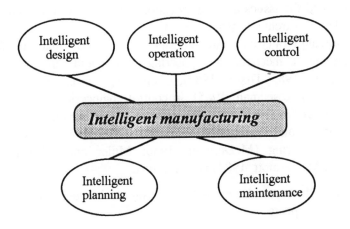

Figure 1.5 Applications of intelligent manufacturing techniques.

(a) Intelligent design

Engineering design is a complicated creative activity. From the viewpoint of design methodology, design is a special process for generating, analyzing and comprehensively evaluating models. The generation of models is a synthesizing process. The evaluation of the models is an analyzing process. As a result, design activities include two kinds of tasks: numerical computation (design analysis) and symbolic reasoning (design synthesis). The current computer-aided design (CAD) technology is very efficient at carrying out the former task, but inefficient for the latter. Nevertheless, design syntheses make a more notable impact on the product quality and manufacturing cost. Therefore, the implementation of design synthesis automation will enhance the quality and efficiency of engineering design. Introducing AI technology into the existing CAD systems is a very effective means of implementing the intelligent design environment that has the capacity for creative design.

Many efforts have been taken to accomplish the objectives above. At the early stage of CAD development, Mann (1965) proposed some conjectures on intelligent design. Brown and Chandrassekaran (1983, 1985) explored the issues of AI applications in engineering design. They studied the hierarchical structure of knowledge representation and problem-solving strategy in routine design. In the 1980s, many intelligent design systems in mechanical engineering (Dixon and Simmons, 1983, 1984; Mostow, 1985; Dixon, 1986), in civil engineering (Sriram *et al.*, 1982; Gero and Rosenman, 1985), in control engineering (James *et al.*, 1985; Rao *et al.*, 1988a and 1988c), in chemical engineering (Stephanopoulos and Davis, 1990; Huang and Fan, 1988), as well as in electrical engineering (McDermott, 1980; Miller and Walker 1988; Bachant and McDermott, 1984) were reported.

Currently, the research on intelligent design techniques consists of the following four aspects:

- Small-scale intelligent design systems in narrow domains. In general, such problems do not involve numerical computation or processing mathematical models. The research objective is to develop intelligent design system prototypes to demonstrate the capacity of AI technology.

- Large-scale coupling intelligent design systems. These intelligent systems combine AI techniques with conventional CAD programs to deal with complicated engineering design problems.

- Intelligent computation and analysis. In engineering design, many commercial software packages for optimization and finite element analysis are very difficult to use. AI technology can improve the efficiency and transparency of these software systems. In addition, many analysis tasks need to use expert's experience and to process symbolic information.

- Intelligent graphics interface. This provides more efficient interpretation and operation to overcome the shortcomings of existing CAD systems and intelligent design systems.

(b) Intelligent operation

The research on intelligent robotics is an important content in CIMS engineering. An intelligent robot should be able to sense (seeing and touching), to think (decision-making), and to act (moving and manipulating). AI technology can help us in dealing with four basic types of robotics problems (Soroka, 1984): kinematics and design, robot selection, workspace layout, planning and maintenance.

Computer-aided process planning (CAPP) intends to build the bridge for implementing the integration of computer-aided product design (CAD) and manufacturing (CAM). Many problems on intelligent CAPP, including representation of geometric and non-geometric features, problem-solving strategies, identification of parameters and specifications, as well as CNC machine programming, have been discussed (Cheung and Dowd, 1988).

Continuous manufacturing process operation can be divided into four situations: startup, normal operation, emergency situation and shutdown. The majority of process automation research and development focus on normal operation that has been very well studied, and run under fully automatic control. However, the other three stages are the most dangerous but are still operated manually, and lack sufficient attention from academic or industrial research (Rao *et al.*, 1991b). Most AI applications in industrial process have been studied for process control or design (Stephanopoulos and Davis, 1990). Process startup automation and process operation support are other areas where intelligent systems could be of value. However, very little previous research has been done in this area. A literature search produced only a few references.

An intelligent process startup automation system, namely IPSAS, was developed (Rao *et al.*, 1991c). The objective of this research was to develop an intelligent system to automate the process startup operation by providing the proper and safe startup procedures. Many injuries and fatalities in the chemical process industries have been due to human errors during the operation of a plant (Kletz, 1988). Proper procedures must be followed during plant operation in order to reduce incidents (Croce *et al.*, 1988). With the use of computer control for plants, some increase in safety is achieved but other hazards arise. This is due to the computer's inability to perform as a human operator. It lacks the intelligence and expertise of the operation experts (Pearson and Brazendale, 1988). Thus, the research project concentrates on the development of an intelligent system to select the proper and safe startup procedures.

(c) Intelligent control

Most of the existing intelligent systems for industrial control are implemented for design purposes (Pang and MacFarlane, 1987). Control system design is a very complex and not a completely understood process, particularly since it is abstract and requires creativity. Such design is largely a trial-and-error process, in which human design experts often employ heuristics or the rules of thumb. An intelligent system can provide a user–computer interface in such a process to improve the efficiency of design (Lamont and Schiller, 1987).

The primary function of the controller can be improved by introducing AI technology into the control system (Astrom, 1985). The experience from building expert systems also shows that the power of expert systems is most apparent when the problem considered is sufficiently complex (Astrom *et al.*, 1986). Moore *et al.* (1984) proposed a real-time expert system for the supervision of a control system, an intelligent fault diagnosis, as well as alarm and performance analysis. Similar investigations were made by Taylor and Frederick (1985) and SRI International (Wright *et al.*, 1986). These expert systems were developed for a single and specific purpose.

Fuzzy set theory was proposed by Zadeh (1978) to represent inherent vagueness. Recently, it has been applied to expert-aided control systems to deal with the imperfect knowledge of control system. For example, a set of fuzzy rules and a fuzzy temporal model are built in an expert system so that the dynamic behavior of the process can be well described (Qian and Lu, 1987).

In the analysis and synthesis of control systems, simulation is a major technique. The traditional simulation techniques are algorithm-based. They are often inflexible and of limited capability to the user. In fact, such techniques cannot clearly simulate the dynamic behavior of the controlled processes. Therefore, the knowledge-based simulation system was suggested to solve the problems encountered above (Zeigler, 1984; Shannon *et al.*, 1985, Shannon, 1986).

A functional approach to designing expert simulation systems for control which perform model generations and simulations was proposed by Lirov and his colleague (1988). They chose the differential games simulator design to build this simulation system. The knowledge representation of the differential games models is described using semantic networks. The model generation methodology is a blend of several problem-solving paradigms, and the hierarchical dynamic goal system construction serves as the basis for model generation. This discrete event approach, based on the geometry of the games, can obtain the solution generally in much shorter time.

(d) Intelligent planning

Production planning in a manufacturing environment is a complex task. Until now, little has been well understood about a production planning process. In addition, production scheduling decisions for large and complex manufacturing facilities are often not made independently by any single individual (Peng *et al.*, 1988). Instead, a group of people who cooperate and share production has to be identified in the organization.

Because of the routine nature of the planning task, the group usually adopts some organization structure to improve the decision-making efficiency. Unfortunately, the process of constructing the organization has not been understood clearly.

Intelligent planning is engaged in understanding the nature of problem-solving by a group of intelligent agents and studying distributed problem-solving and planning strategies. For example, Fikes (1982) studied informal cooperative work in an office setting, and modeled an agent's work in terms of commitments to other agents.

The research on parallel processing between organization theory and the design of distributed problem solving was explored (Fox, 1984). Fox and Smith (1984) pointed out that such factors as task complexity, uncertainty and scarcity of resources were important in deciding how to distribute a system.

Intelligent planning for a flexible manufacturing system (FMS) aims at implementing the automation of production scheduling in the CIM environment through linking an information management system with a material manufacturing environment. Many scientific papers and industrial applications have been presented. A constraint-based framework OPIS (Smith, 1987) for reactive management of factory schedules was developed. Kim *et al.* (1988) developed a knowledge acquisition environment for FMS scheduling and proposed the knowledge base organization and knowledge acquisition approach. Other early applications of intelligent planning systems can be found in the literature (Parrello *et al.*, 1986; Rit, 1986; Steffen and Greene, 1986; Stefik, 1981). The design of intelligent planning systems was discussed in detail (Rickel, 1988).

(e) Intelligent maintenance

Equipment maintenance involves a diagnostic procedure incorporating many rules as well as judgment decisions. Experts' knowledge and experience are very important factors when an engineer locates a failure problem and implements an appropriate correction.

There exist many intelligent systems developed for fault diagnosis.

Such an application ideally lends itself to rule-based procedures (Miller and Walker, 1988). Intelligent maintenance systems are now being utilized to assist maintenance personnel in performing complex repairs by presenting menu-driven instruction guides for diagnosing, interpreting, predicting and monitoring equipment faults.

The intelligent maintenance system is one of the most successful AI applications in manufacturing. Many applications have been reported. General Motors uses an intelligent system, built using S.1™ (built by Teknowledge™), to increase the accuracy and efficiency of service technicians at over 10 000 independent dealerships (Miller and Walker, 1988).

Another successful example is ASCIT (all-purpose, simple-to-use, computer-based interactive tool) (Miller and Walker, 1988). It can facilitate maintenance and diagnostics of any equipment or machinery. ASCIT provides users access to system information which includes textual descriptions, visual information (picture, drawing, and schematic), and step-by-step audio and visual information on preventive, troubleshooting and corrective maintenance.

An integrated distributed intelligent system for large mining trucks (240 tons) condition monitoring and maintenance has been developed (Rao *et al.*, 1992b). The system can assist a manager in setting production and scheduling maintenance strategies for these vehicles, analyzing the data from sensors, and processing the information with production statistic and system analysis models as well as domain-specific knowledge. Finally, it provides suggestions and measures for better truck maintenance.

According to the summary of the preview research projects, intelligent systems for manufacturing industries can be divided into four types.

- **Type 1.** Single symbolic reasoning system that only processes symbolic information, and provides assistance to engineers in a decision-making process for manufacturing.

- **Type 2.** Coupling intelligent system that links numerical computation programs with a symbolic reasoning system such that it can be used to solve engineering problems.

- **Type 3.** Artificial neural networks that model the human brain and deal with empirical data.

- **Type 4.** Integrated distributed intelligent system that is a large-scale intelligent integrated environment. It can integrate different expert systems, numerical programs, neural networks, as well as computer graphics packages to solve complex engineering problems.

1.2.4 Symbolic reasoning systems

Symbolic reasoning systems are developed to solve ill-structured problems that are difficult to handle by purely algorithmic methods. The knowledge base of a symbolic reasoning system contains two kinds of knowledge: public knowledge and private knowledge (expertise). It is a specific computer program, implemented by a specific programming technique, and used to solve a specific problem in a specific domain. The segregation of the database, the knowledge base and the inference engine in the expert system allows us to organize the different models and domain expertise efficiently because each of these components can be designed and modified separately.

Presently, symbolic reasoning systems are extensively applied in the research of intelligent manufacturing systems. Among the successful AI applications, most intelligent systems are production systems (Talukdar *et al.*, 1986). Production systems facilitate the representation of heuristic reasoning such that intelligent systems can be built incrementally as the expertise knowledge increases.

In our early projects, OPS5 (Brownston *et al.*, 1985) served as one of the programming tools, which was the most widely used language for developing symbolic reasoning systems based on production rules. The tool consists of three components: a database (working memory), a knowledge base (production memory) and an inference engine as shown in Figure 1.6. Working memory is a special buffer-like data structure and holds the knowledge that is accessible to the entire system. Each unit of working memory is an attribute-value element. Any attribute that is not assigned a value for a particular instance is given the default value designated as 'nil'.

Production memory contains the general knowledge about a problem. The expert knowledge of the problem is described by a set of production rules stored in production memory. The typical production rule is described as 'IF (condition), THEN (action)'. Every production rule consists of a condition–action pair. In OPS5, the condition part is called the 'LEFT-HAND-SIDE (LHS)', and the action part is called the 'RIGHT-HAND-SIDE (RHS)'. Each condition element specifies a pattern that is to be matched against working memory. The matching process will be described as we define the syntax of LHS. The actions forming RHS are imperative statements that are executed in sequence when the rule is fired. Most permissible actions alter the working memory.

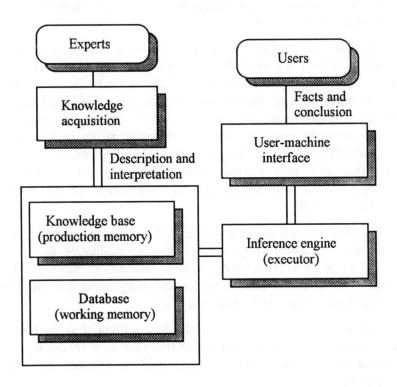

Figure 1.6 Structure of single symbolic reasoning systems.

The inference engine is an executor. It must determine which rules are relevant to a given data memory configuration and select one to apply. Usually, this control strategy is called conflict resolution. The inference engine can be described as a finite-state machine with a cycle consisting of three steps; that is, matching rules, selecting rules and executing rules.

1.2.5 Coupling intelligent systems

Many existing symbolic reasoning systems are developed for specific purposes, which are implemented with a Lisp-like language, and production rules are used to represent domain expertise. In light of applications, these systems can only process symbolic information and make heuristic inferences. The lack of numerical computation and coordination between single applications limits their capability to solve real engineering problems.

Historically, computer-aided manufacturing techniques, such as CAD/CAM/CAE, followed the development of mathematical computing theory and techniques. Some reports stated that mathematical modeling is not the only means to describe real manufacturing problems (Rao *et al.*, 1988a). The knowledge and experience we know about the world are not well captured by numbers, since our reasoning is not well modeled by arithmetic (Davis, 1987). Many manufacturing problems are ill-structured ones that deal with non-numerical information and non-algorithmic procedures, and are suitable for the application of AI techniques (Buchanan, 1985).

We often use qualitative and quantitative analyses together in solving engineering problems (Figure 1.7). Usually, qualitative decisions are mainly based on symbolic and graphical information, while quantitative analysis is more conveniently performed using numerical information. Both methods often complement each other.

Any numerical solution, no matter how perfect it is, is always an approximation to the true solution. The true solution is always represented analytically. Analytical solutions can only be obtained by symbolic processing.

A main disadvantage of the existing symbolic reasoning systems is their inability to handle numerical computation. This makes these systems less useful for many complicated engineering problems. In a manufacturing environment, we require not only a qualitative description of system behavior, but also a quantitative analysis. The former can predict the trend of the change of an operating variable, while the latter may provide us with a means to identify the change range of the variable.

Moreover, as a part of the accumulated knowledge of human expertise, many practical and successful numerical computation packages have been made available. We agree that artificial intelligence should emphasize symbolic processing and non-algorithmic inference (Buchanan, 1985), but it should be noted that the utilization of numerical computation will make intelligent systems more powerful in dealing with engineering problems. Like many modern developments, artificial intelligence and its applications should be viewed as a welcome addition to the technology, but they cannot be used as a substitute for numerical computation.

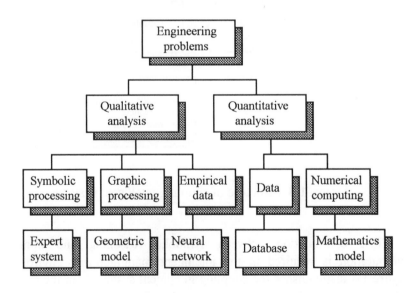

Figure 1.7 Qualitative and quantitative analyses.

The coordination of symbolic reasoning and numerical computation is essential to developing intelligent manufacturing systems. More and more, the importance of coordinating symbolic reasoning and numerical computing in knowledge-based systems is being recognized. It has been realized that if applied separately, neither symbolic reasoning nor numeric computing can successfully address all problems in manufacturing. The complicated problems cannot be solved by purely symbolic or numerical techniques (Jacobstein *et al.*, 1988; Wong *et al.*, 1988).

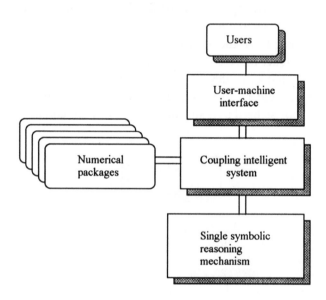

Figure 1.8 Structure of coupling intelligent systems.

Close coordination between symbolic reasoning and numerical computation is required in the intelligent manufacturing environment. Figure 1.8 distinguishes the coupling intelligent system from the symbolic reasoning system from the viewpoint of software architecture. So far, many coupling intelligent systems have been developed in various engineering fields to enhance the problem-solving capacity of the existing symbolic reasoning systems (Kitzmiller and Kowalik, 1987). In IDSOC (intelligent decisionmaker for problem-solving strategy of optimal control) (Rao *et al.*, 1988c), a set of numerical algorithms to compute certainty factors is coupled in the process of symbolic reasoning. Another coupling intelligent system SFPACK (Pang *et al.*, 1990) incorporates expert system techniques in design package then supports more functions to designers. Written in Franz Lisp, CACE-III (James *et al.*, 1985) can control the startup of several numerical routines programmed in FORTRAN. Similar consideration was taken into account by Astrom's group (Astrom *et al.*, 1986). More and more, the coordination of symbolic reasoning and numerical computing in knowledge-based systems attracts much attention.

The methods of integrating single symbolic reasoning systems and numerical computation packages were proposed by Kitzmiller and Kowalik (1987). A few developers tried to develop coupling intelligent systems with conventional languages, such as FORTRAN, so that these symbolic reasoning systems could be used as subroutines in a FORTRAN main program. Others suggested developing coupling intelligent systems in conventional languages in order to achieve the integration of numerical algorithms and symbolic inference. However, these methods prohibit developing and using the individual programs separately. This makes acquiring new programs very difficult, and is not cost-effective. Another disadvantage is that the procedural language environment cannot provide the many good features that Lisp provides, such as easy debugging and allowance for interruption by human experts.

Numerical languages often have a procedural flavor, in which the program control is command-driven. They are very inefficient when dealing with processing strings. Symbolic languages are more declarative and data-driven. However, symbolic languages are very slow to execute numerical computations. Coupling of symbolic processing with numerical computing is desirable to use numerical and symbolic languages in different portions of a software system.

Currently, not all of the expert system tools or environments provide programming techniques for developing coupling intelligent systems. For example, it is very difficult to carry out numerical computation in OPS5 (Brownston *et al.*, 1985; Rao *et al.*, 1988a). However, many software engineers are now building the general-purpose tools for coupling intelligent systems that will be beneficial to artificial intelligence applications to engineering domains.

1.2.6 Artificial neural networks

The human brain is an example of a natural neural network. Neurons are living nerve cells, and neural networks are networks of these cells. Such a network of neurons can think, feel, learn and remember. Artificial neural networks (ANNs) are modeled after the known capabilities of the human brain are copied into computer hardware and software (Vemuri, 1988). ANNs exhibit the following characteristics:

- Learning: ANNs can modify their own behaviors in response to their environment.

- Generalization: Once trained, ANN responses can be insensitive to minor variations in their inputs.

- Abstraction: Some ANNs are capable of abstracting the essence of a set of inputs.

- Applicability: ANNs become the preferred technique for a large class of pattern recognition tasks that conventional computers do poorly.

The artificial neuron is designed to mimic the first-order characteristic of the biological neuron. A set of inputs is applied to a neuron. Each input, representing the output of another neuron in the previous layer, is multiplied by an associated weight, and all of the weighted inputs are then summed to produce a signal. This signal is usually further processed by an activation function to produce the neuron's output. This activation function simulates the nonlinear transfer characteristics of the biological neuron, and is often chosen to be the sigmoidal logistic function or the hyperbolic tangent function. The output of the neuron is then distributed to other neurons in the next layer.

Neural network structures can be divided into two classes based upon their operations during the recall phase: feedforward and feedback. In a feedforward network, the output of any given neuron cannot be fed back to itself directly or indirectly, and so its present output does not influence future output. Grandmother cells, ADALIN, MADALINE and backpropagation networks are feedforward networks. In a feedback network, any given neuron is allowed to influence its own output directly through self-feedback or indirectly through the influence it has on other neurons from which it receives inputs. Kohonen self-organizing, and Hopfield networks are feedback networks.

A network is so trained that application of a set of inputs produces the desired (or at least consistent) set of outputs. Each input (or output) set is referred to as a vector. Training is accomplished by sequentially applying input vectors, while adjusting network weights according to a predetermined procedure. During training, the network weights gradually converge to values such that each input vector produces the desired output vector. Training algorithms are categorized as supervised and unsupervised. Supervised training requires the pairing of each input vector with a target vector representing the desired output. An unsupervised training algorithm requires no target vector for the output, and hence, modifies network weights to produce output vectors that are consistent.

Recently, the interest in ANNs has been growing rapidly since neural networks hold the promise of solving problems that have proven to be extremely difficult for traditional numerical computation. There are varieties of ANN algorithms available. Each of them has specific advantages for some kinds of problems. Neural network problems can be categorized into the following groups based on what specific information we want to give the network as input and what we expect the network's output to be.

- Mapping: In a mapping problem, an input pattern is associated with a particular output pattern.

- Associative memory: An associative memory stores information by associating it with other information. The recall is performed by providing the association and having the network produce the stored information.

- Categorization: The network is provided with an input pattern and responds to the category to which the pattern belongs.

• Temporary mapping: Such a mapping includes the consideration of time. Many process control patterns are temporal mapping problems.

ANNs have been applied to many pattern recognition problems including image processing, speech recognition, sensor interpretation, and motor control. In the field of manufacturing engineering, ANNs appear to be particularly suitable for fault detection and diagnosis (Watanabe *et al.*, 1989), process control (Donat *et al.*, 1990), modeling and simulation.

1.2.7 Development of intelligent systems

Figure 1.9 shows the basic stages for developing an intelligent system. The development process is divided into five stages. At the first stage, i.e. **identification**, the following issues have to be considered: (i) problems (What kind of problems to be solved? How to define these problems? Is human expertise needed to solve the problems?); (ii) participants (domain experts, end-users and intelligence engineers); (iii) resources (knowledge resource, development time, computer facility and software tools); (iv) goals (to solve problems that conventional programming techniques cannot address, to improve the efficiency of decision process, and to deal with the specific applications with ambiguous, uncertain or complex subjects).

At the **conceptualization** stage, the problem is decomposed into sub-problems, knowledge acquisitions are performed, and input/output relationships are analyzed. The **formulization** process involves mapping the key concepts, sub-problems and information flow characteristics isolated in the conceptualization stage into the formal representation. Knowledge is analyzed, and the user-interface is designed. The fourth stage (**implementation**) transfers the formalized knowledge into the representation framework associated with the development tools chosen for this problem. The final stage (**evaluation**) involves testing and modifying the prototype intelligent system and its representational forms used to implement the system.

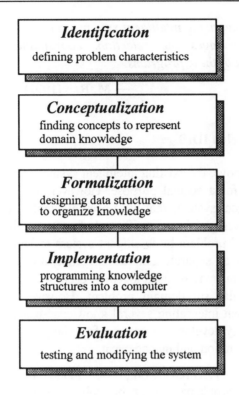

Figure 1.9 Intelligent system development stages.

1.3 Integrated distributed intelligent systems

Integrated Distributed Intelligent System is a large-scale knowledge integration environment, which consists of several symbolic reasoning systems, numerical computation packages, neural networks, database management subsystem, computer graphics programs, an intelligent multimedia interface and a meta-system. The integrated software environment allows the running of programs written in different languages and the communication among the programs as well as the exchange of data between programs and database. These isolated intelligent systems, numerical packages and models are under the control of a supervising intelligent system, namely, the meta-system. The meta-system manages the selection, coordination, operation and communication of these programs.

— **M. Rao (1991)** —

1.3.1 Why develop IDIS

As discussed above, the existing intelligent systems can only be used alone and inflexibly for a special purpose. However, we cannot integrate the isolated intelligent systems that have been available, even though each of them is well developed for a specific task. In general, the best way to solve this complicated problem by intelligent system techniques is to distribute knowledge and to separate domain expertise. In such a case, several intelligent systems may be used together, each of them being developed to solve a particular sub-domain problem. However, current technology prohibits us from integrating several knowledge-based systems that have been successfully developed. Here, we face the problem of heterogeneous knowledge integration and management.

In previous research work, we found that many industrial manufacturing problems cannot be solved using the existing expert systems or numerical computation techniques. Many process variables that affect process operation and product quality and quantity are only partially understood, and sometimes may not be directly measurable. For example, some empirical data from manufacturing cannot be effectively processed using existing expert systems or numerical computation packages. A new methodology integrating neural networks, symbolic reasoning, numerical computation as well as graphic simulation may be suitable to deal with such data.

In manufacturing industry, a vast mount of operation data and information from various sensors will be processed in time. So far, database technology and numerical computation methods have been used to deal with operation data. However, how to acquire knowledge effectively from the data source still remains a difficult issue.

Conflicts usually arise because different domain expert systems may make different decisions on the same issues based on different criteria. Most intelligent systems for industrial manufacturing processes work in an interdisciplinary field, and often deal with conflicting requirements. These intelligent systems may produce conflict decisions even based on the same information. One of the bottlenecks for applying intelligent systems commercially is to integrate different quantitative and qualitative methods.

Distributed intelligent system technology (Bridgeland and Huhns, 1990; Conry *et al.*, 1990) is a very practical method for improving the ability of knowledge base management and maintenance. Since most manufacturing problems need knowledge and experience from different areas, the integration of distributed intelligent systems is a significant factor in solving more complicated manufacturing problems.

The modern industrial manufacturing process is becoming increasingly difficult for operators to understand and operate effectively. Due to the importance of the operator's role in these systems, the quality of interaction between human operators and computers is crucial. Thus, developing an intelligent multimedia interface, which communicates with operators via multiple media and modes such as language, graphics, animation and video (Maybury, 1992), will facilitate human operators interaction with complex real-time monitoring and control systems.

Besides the above applications, the following drawbacks often arise in existing intelligent manufacturing systems:

- lack of efficient search methods to process different knowledge in large decision-making space. For example, in the conceptual design of chemical processes, there is a very large number of alternatives (10^7) to be considered to accomplish a design task (Douglas, 1988),
- lack of coordination of symbolic reasoning, neural networks, numeric computation, as well as graphics representation,
- lack of integration of different intelligent systems, software packages and commercial AI tools,
- lack of efficient management of intelligent systems,
- lack of capability to handle conflicts among intelligent systems,

- lack of a parallel configuration to deal with a multiplicity of knowledge representation and problem solving strategies, and

- difficulty in modifying knowledge bases by end-users rather than the original developers.

In solving the problems mentioned above, the key issue is how to develop a meta-system and implement multiple media integration. Concepts of integrated distributed intelligent system were proposed by Rao *et al.* (1987b). In the past several years, the Intelligence Engineering Laboratory at the University of Alberta has been engaged in studying and developing a new architecture to control and manage large-scale intelligent systems for industrial manufacturing applications. So far, a prototype integrated distributed intelligent system (IDIS) platform has been implemented.

IDIS is a large knowledge integration environment, which consists of several symbolic reasoning systems, numerical computation programs, database management subsystem, computer graphics packages, multimedia interfaces, as well as a meta-system. It makes use of the advanced object-oriented programming technique in C++ language. The integrated software environment allows the running of programs written in different languages, communication among programs, as well as the exchange of data between programs and databases. These isolated expert systems, numerical packages and programs are under the control of a supervising expert system, namely the meta-system. The meta-system manages the selection, coordination, operation and communication of these programs. The integrated distributed intelligent system is illustrated in Figure 1.10.

1.3.2 Advantages of IDIS

An integrated distributed intelligent system has the following advantages:

- provides an open architecture to help users protect their investment when introducing new computer technology;

- coordinates all symbolic reasoning systems, numerical computation routines, neural networks, and computer graphics programs in an integrated environment;

- distributes knowledge into separate expert systems, numeric routines and neural networks so that the knowledge bases of these systems are easier to change by commercial users other than their original developers;

- acquires new knowledge efficiently;

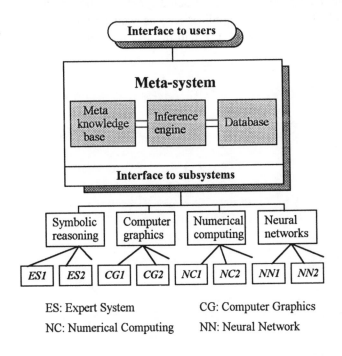

ES: Expert System   CG: Computer Graphics
NC: Numerical Computing   NN: Neural Network

Figure 1.10 Functions of integrated distributed intelligent system.

- finds a near optimal solution for conflicts and facts among different intelligent systems;

- provides the capability of parallel processing in an integrated distributed intelligent system; and

- communicates with measuring devices and final control elements in the control systems, and transforms various input/output signals into the standard communication formats.

Integrated distributed intelligent system (IDIS) is a key technology to realize the manufacturing automation at the knowledge-intensive stage. This new software integration platform can utilize different knowledge (analytical and heuristic knowledge), process different information (symbolic, numerical, and graphic information), and integrate different computer languages. It has attracted great attention from both academia and industry. IDIS, its applications, as well as the meta-system are the main topics in this book. We will discuss them in detail in the following chapters.

2

Meta-system architecture

2.1 Meta-system

2.1.1 Integration of intelligent systems

So far, many intelligent systems have been developed. However, their capability of dealing with complicated manufacturing problems is very limited. The main disadvantages of the existing intelligent systems have been summarized in section 1.3.1. The two most urgent problems to be solved are how to coordinate symbolic reasoning, numerical computation, neural networks, as well as computer graphics, and how to integrate heterogeneous intelligent systems.

With the accumulated experience of building expert systems, we have realized that integrating different intelligent systems into a large-scale knowledge environment is often necessary but difficult. In such an environment, a supervisory system that controls and manages the heterogeneous intelligent subsystems is required. The supervisory system has to provide integration functions in the following phases:

(1) integration of knowledge of different disciplinary domains;

(2) integration of empirical expertise and analytical knowledge;

(3) integration of various objectives, such as research and development, system design and implementation, process operation and control;

(4) integration of different symbolic processing systems (expert systems),

(5) integration of different numerical computation packages,

(6) integration of symbolic processing systems, numerical computation systems, neural networks, and computer graphics packages,

(7) integration of multimedia information, such as symbolic, numerical and graphic information.

Phases 1 and 2 are at the knowledge level. Phase 1 also indicates the characteristics of modern engineering techniques. In our case, it means the integration of such heterogeneous knowledge as that about marketing, management, manufacturing, system engineering and computer science, etc. (see Figure 1.1). Phase 3 functions at both the knowledge and the functional levels. Phases 4 through 7 perform their integration functions at the functional level, through the problem-solving level to the program level.

2.1.2 Integration of manufacturing activities

The industrial manufacturing process includes two stages: manufacturing and testing. These two stages are operated based on the information provided in the design stage. There are two flows in the process: material flow and information flow. Material flow consists of all necessary working stations where workpieces are processed, and the transportation of workpieces between these stations. Information flow can be divided into two branches based on functions: domain information flow (or technical information flow), and meta-information flow (or decision-making information flow). Domain information flow carries the information about product design, manufacturing and testing. Meta-information flow describes system integration, knowledge management and coordination in manufacturing processes. Meta-information flow controls domain information flow and material flow. These three flows compensate each other and coexist to form a complete system. In such a way, the integration environment for distributed intelligent system is needed.

2.1.3 Meta-system

For the sake of convenience, a frame representation technique is applied to describe the hierarchical architecture of the integrated distributed intelligent system. The hierarchical structure of the software environment is demonstrated in Figure 2.1. By integrating distributed intelligent systems, we can achieve efficiency as well as conceptual and structural advantages. This new conceptual design framework can serve as a universal configuration to develop high-performance intelligent systems for many complicated applications.

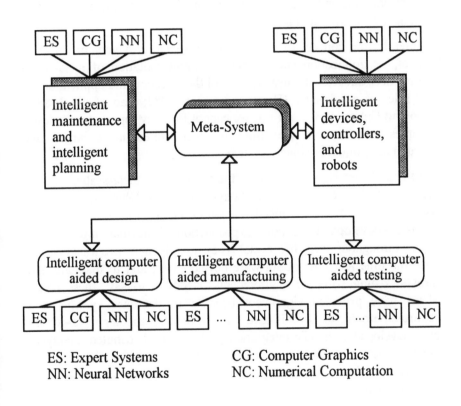

ES: Expert Systems   CG: Computer Graphics
NN: Neural Networks   NC: Numerical Computation

Figure 2.1 Integration of manufacturing software systems.

The key issue in constructing an integrated distributed intelligent system is to organize a meta-system. The meta-system can thus be referred to as a 'control mechanism of meta-level knowledge' (Rao, *et al.*, 1987b). Knowledge can be categorized into 'domain knowledge' and 'meta-knowledge'. The former is usually defined as facts, laws, formulae, heuristics and rules in specific domains of knowledge, whereas the latter is defined as the knowledge about domain knowledge (Davis and Lenat, 1982), and can be used to manage, control and utilize domain knowledge. Meta-knowledge possesses diversity, covering a broader area in content and varying considerably in nature. Also, it has a fuzzy property. Hence, the representation and knowledge base organization of meta-knowledge should have their own characteristics.

Presently, the concept of meta-level knowledge is a very broad one. However, its application concentrates on implementation of meta-rules that have been successfully used to control the selection and application of object-rules or a rule-base (Davis and Lenat, 1982; Rao *et al.*, 1988c; Orelup and Cohen, 1988). As yet, we cannot find any reported meta-level techniques which solve the problems encountered in a large-scale integrated distributed intelligent system.

In the software engineering field, the terminology meta-system is also widely used. For example, a meta-system for software specification environments has been developed (Dedourek *et al.*, 1989). These meta-systems are developed to support the production of information processing systems throughout their life cycle. They greatly help programmers to reduce the time and cost of software development, and to maintain and improve existing software environments. Needless to say, these meta-systems have provided significant benefits to database systems and software environments. In this text, we will discuss another meta-system that is developed to handle integration problems in distributed intelligent systems. Such a meta-system is different from those discussed by Dedourek and his colleagues (1989) in both aspects of concept and implementation.

Not just extending the concept of meta-level knowledge, but rather proposing an innovative idea, we develop a new high-level supervisory system, that is, a meta-system, to control IDIS.

2.2 Meta-system functions

The main functions of a meta-system are described as follows:

2.2.1 Coordination function

The meta-system is the coordinator to manage all symbolic reasoning systems, numeric computation routines, neural networks, and computer graphics programs in an integrated distributed intelligent system.

The hierarchy of our integrated distributed intelligent system as described in Figure 2.2 indicates that different types of information (symbolic, graphic and numerical) are utilized and processed together.

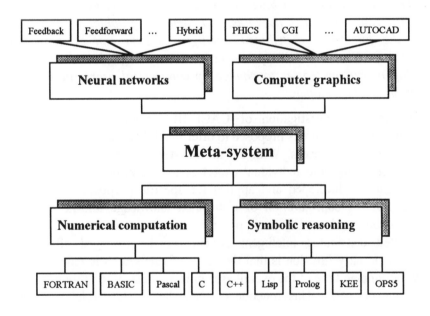

Figure 2.2 Software structure of integrated distributed intelligent system.

Even in the branch of symbolic processing, many different languages and tools (such as Lisp™, Prolog™, OPS5™, KEE™, C++™, and so on) may be applied to build individual symbolic reasoning systems. In each language family, there also exist different expert systems. For example, among the OPS5™ family, there are IDSOC and IDSCA (Rao *et al.*, 1988a and 1988c). The meta-system controls the selection and operation of all programs in an integrated distributed intelligent system, and executes the translation between different data or programs. For example, it may translate a set of numerical data into symbolic sentences, and then send the symbolic information to an expert system that is to be invoked next.

2.2.2 Distribution function

The meta-system distributes knowledge to separate expert systems, numeric routines, neural networks, and computer graphics programs so that the integrated distributed intelligent system can be managed effectively. Such modularity makes the knowledge bases of these intelligent systems easier to change by commercial users rather than their original developers.

Expanding the investigation of CACE-III (James *et al.*, 1985), a distributed architecture is proposed to offer flexibility and efficiency in the decision-making process. With the meta-system, the knowledge sources can be separated into individual expert systems and numeric programs, which can be developed and applied for specific purposes at different times. Such a configuration of distributed knowledge allows users to write, debug and modify each program separately so that the overall integrated distributed intelligent system can be managed efficiently. Another primary function is that the scope of rule search is reduced such that the running time of the overall system is minimized, to accomplish the real-time objective. This function can also help reduce rule or system interaction.

The basic approach is to distribute the domain knowledge for a complex problem into a group of expert systems, numeric routines, neural networks, as well as computer graphics, then to attempt to limit and specify the information flow among these programs. This approach makes the integrated intelligent environment much easier to handle.

2.2.3 Integration function

The meta-system is the integrator which can help us easily acquire new knowledge.

The meta-system provides us with a free hand in integrating and utilizing new knowledge. Each of the programs in an integrated distributed intelligent system is separated from every other, and only ordered strictly by the meta-system. Any communication between two programs must rely on translation by the meta-system. This configuration enables us to add or delete programs much more easily. When a new expert system or numerical package is to be integrated into the integrated distributed intelligent system, we just have to modify the interface and the rule-base of the meta-system, while other programs are kept unchanged. The most significant fact is that successfully developed software packages can be applied wherever needed. The time and money required in doing 'repeat work' can be saved, making the commercial application of intelligent systems feasible.

2.2.4 Conflicting reasoning function

The meta-system can provide a near optimal solution for conflicting solutions and facts among different intelligent systems.

Knowledge sources are distributed into various distinctive programs that correspond either to different tasks or to different procedures. However, conflict usually arises because different domain expert systems can make different decisions based on different criteria, even though the data that fire the rules are nearly the same. In particular, intelligent manufacturing is interdisciplinary in nature. It allows the knowledge from engineering, artificial intelligence and computer science to be extensively applied to manufacturing processes. Therefore, IDIS often deals with conflicting requirements, in which various domain knowledges are utilized and different expert systems are developed. In this case, the meta-system should be able to pick a near optimal solution from the conflicting facts when the requirements from different domains contradict each other. This function will play an even more important role in knowledge integration and management in the future.

Currently, a method has been developed to search for a near optimal solution in a conflicting decision-making process, which is based on priority ranking; that is, different objectives (such as criteria, facts, and methods) are assigned different priority factors, thus each solution will result in an overall priority factor. This is similar to that in certainty factor calculation. However, this method relies greatly on the expertise of the person who ranks the priority in terms of various objectives. It may not be easy to use for general cases.

2.2.5 Parallel processing function

The meta-system provides the possibility of parallel processing in an integrated distributed intelligent system.

In a production system, all rules and data are effectively scanned in parallel at each step to determine which rules are triggered on which data. This inherently parallel operation can perform very well in the integrated distributed intelligent system (Okuno *et al.*, 1988).

The one-to-one connection between a program and the meta-system allows the execution of two or more individual programs at the same time since, in a parallel computer, several inference engines are able to deal with simultaneous bidirectional resolution (Handelman and Stengel, 1987). For example, forward-chaining can be employed on one processor and backward-chaining on another. This allows dynamic resolution strategies (Burg *et al.*, 1985). The execution of a rule in the meta-system may result in several different actions that can invoke two or more programs to operate separately at the same time. This function can greatly enhance the quality of real-time computation. Also, it may be applied to other computational processes.

2.2.6 Communication function

The meta-system can communicate with the measuring devices and the final control elements in the control systems and transform various non-standard input/output signals into standard communication formats.

A real-time manufacturing environment always deals with the transformation of communication signals that are either inputs or outputs, such as electronic (current, voltage), pneumatic, acoustic or even optical signals, on different scales. The meta-system provides a user-friendly interface to manufacturing systems so that these signals can be exchanged, depending on the needs of the users or other intelligent systems. For example, a rise in the output water temperature of a boiler results in an increase of electronic current from a thermocouple. Through the interface transformation of the meta-system, the signal may be expressed by a numerical value or symbolic information, and be used for computing or reasoning. A corresponding signal will be produced once a result is obtained in an integrated distributed intelligent system, and the signal can be used to manipulate a final control element, such as a control valve.

In order to illustrate the function of a meta-system, let us visualize an integrated distributed intelligent system as a business company. The meta-system acts as a manager who assigns different jobs to employees according to their expertise, and monitors the progress of their work, as indicated by the coordination and distribution functions. When a newly-hired employee joins the company, the manager is responsible for providing the new job assignment, and possibly for modifying the work plan. Meanwhile, other people still do essentially the same jobs as they did before. This simulates the integration function. If the suggestions from two employees are in conflict, the manager must decide on a near optimal solution between them, similar to the conflicting reasoning function of the meta-system. The parallel processing function can be visualized as a situation where the manager can let two employees work on their own projects at the same time but share the same tools and resources. Of course, the function of parallel processing is more sophisticated than the simple management system described here. It requires further investigation and study. Finally, the transformation between production plan and market information demonstrates the communication function.

2.3 Configuration of meta-system

Like the common expert systems, the meta-system has its own database, knowledge-base and inference engine, but it distributes its activities into the separated, strictly ordered phases of information gathering and processing. The meta-system configuration as shown in Figure 2.3, includes the following six main components: an interface to the external environment, an interface to internal subsystems, a meta-knowledge base, a global database, a static blackboard, and an inference mechanism.

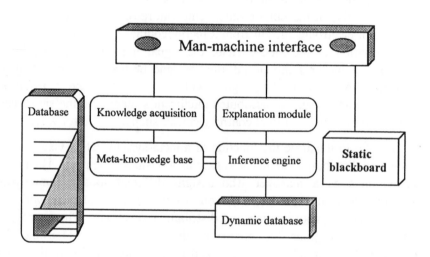

Figure 2.3 Configuration of the meta-system.

2.3.1 Interface to external environment

The interface to the external environment builds the communication between the users and internal software systems as well as among the external software systems. The interface includes an icon-structured menu that consists of windows, data structure, security module, as well as editor module. Windows display different media of information. Data structure is used to receive information. A security module identifies the classes of users. The editor module helps users access all knowledge bases and databases at meta-level and subsystem-level. Furthermore, the interface provides more intelligent functionality such as natural language recognition for oral inputs and computer vision for handwriting inputs. The interface plays a key role in an open structured software system in two ways: the first one is to codify human expertise into the computer system such that it can adopt the most creative intelligence and knowledge in decision-making. Another is to communicate with other intelligent software systems to extend the system into a much larger scale for more complicated tasks.

2.3.2 Meta-knowledge base

The meta-knowledge base is the intelligence resource of the meta-system. It serves as the foundation for the meta-system to carry out the managerial tasks. The meta-knowledge base consists of a compiler and a structured frame knowledge representation facility. The compiler converts the external knowledge representation, which is obtained through the editor and is easy to understand by users, into internal representation forms available in the inference mechanism. The structure of knowledge representation can be production rule or frame or their combination. Characterized by diversity and variety in nature, the meta-knowledge may be better represented in object-oriented *frame* structures.

There are several modules in the frame to represent different components of meta-knowledge. These components are functioned for specific purposes. For example, the communication standardization module for heterogeneous subsystems and the conflicting resolution module are formed for a general management purpose at supervisory level. The module for knowledge about subsystems and the task assignment module have to be built according to each specific problem. The meta-knowledge base employs an open organization structure. It allows new intelligent functionality to enter the meta-knowledge base to engage more duties in decision-making.

2.3.3 Global database

The database in the meta-system functions as a global database for the integrated distributed intelligent system, distinguished from the databases in the subsystems, which are attached only to their individual subsystems. The interface converts the external data representation form into an internal form. The data flow in the global database is controlled either by the inference engine, depending on the corresponding module in the meta-knowledge base, or by users of certain security classes.

2.3.4 Inference mechanism

Due to the diversity of the meta-knowledge and the variety of its representation forms, the inference mechanism in the meta-system adopts various inference methods, such as *forward chaining, backward chaining, certain reasoning, uncertain reasoning, conflict reasoning,* and so on. The inference mechanism performs operations and processing on the meta-knowledge. Additionally, it also carries out various actions based on the reasoning results; that is, passing data between any two subsystems, or storing new data in the database. Therefore, there are some functional modules in the mechanism, which further extends the functionality of the inference mechanism.

2.3.5 Static blackboard

The static blackboard is an external memory for temporary storage of information that is needed when the system is running. Limited by the on-board memory space, the subsystems in IDIS are unable to execute at the same time. In fact, it is unnecessary to run the entire system simultaneously. Very often, the meta-system and all subsystems are run on the distributed hardware environment so that there must be a buffer area in the external memory for any two subsystems to exchange information. Besides data storage, the conversion of data in heterogeneous languages into exchangeable standard form is also completed in the static blackboard.

2.3.6 Interface to internal subsystems

This component of the meta-system is established based on each specific application. The internal interface connects any individual subsystems which are used in problem-solving and under the control and management of the meta-system. Each module of the interface converts a nonstandard data form from a specific subsystem into a standard form in the integrated distributed intelligence environment. Conversion between the standard forms of different languages is carried out by the meta-system.

2.4 Meta-system implementation

2.4.1 Implementation in OPS5 environment

The meta-system was first implemented with OPS5™ on a VAX 11/780 computer, running under the UNIX™ operating system. OPS5™ is a Lisp-based tool (Brownston *et al.*, 1985). In Lisp™ language, there is no difference between numerical data and symbolic attributes. A message can thus be an arbitrary expression. Lisp™ programming environment supports good debugging features. OPS5™ was developed at the Carnegie-Mellon University, and is the most popular tool to develop production rule-based systems. Various versions of the OPS5™ expert system development tools are now available for many different computers. For example, OPS5+™, for IBM PC, Macintosh, as well as Apollo Workstation, is available from Computer Thought Company, Plano, TX. OPS5™ supports a *forward chaining inference* process. Its pattern matching methodology permits variable bindings. However, it does not provide facilities for sophisticated object representation, and has difficulty with numerical computation. It is not an easy tool for the non-programmer to use.

In order to meet the requirements for integrated distributed intelligent system, we have to make some modifications to OPS5™. The main tasks here are (i) to increase numerical computation power to the extent that it is capable of solving general engineering problems; (ii) to enable OPS5™ to initialize a sub-process (i.e. to enable another expert system to work), as well as to terminate its sub-processes; (iii) to constantly monitor the progress of its sub-processes.

OPS5™ does not possess enough numerical operations for engineering applications. Actually, the commonly used programming language to deal with numerical computation in manufacturing systems is FORTRAN™. However, OPS5™ is built on the top of Lisp™. Most Lisp™ functions can be used in the OPS5™ environment through the *external function* declaration. In Lisp™, a compiled FORTRAN™ function can be called through a *special loader*. By a simple interface, we can call a FORTRAN™ subroutine into the OPS5™ environment. This procedure is shown in Figure 2.4. Besides FORTRAN™, C™ and Pascal™ can also be utilized in this manner. Of course, certain disciplines have to be obeyed in order to use these foreign functions. However, to combine FORTRAN™ code into the OPS5™ environment is a great challenge, since any FORTRAN™ subroutine can only be called by running a FORTRAN™ main program, and this restricts the use of FORTRAN™ subroutines within Lisp functions. Fortunately, the configuration of the integrated distributed intelligent system allows user to run FORTRAN™ subroutines as separate processes, and then to build communications between FORTRAN™ and Lisp™ through data files and the interface rule-base of the meta-system. Through Lisp™ external functions, the meta-system that is running under the OPS5™ environment can call a numerical computation routine into the integrated distributed intelligent system.

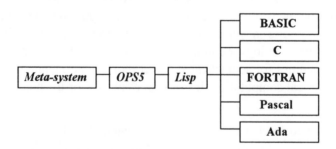

Figure 2.4 Calling heterogeneous language programs.

Once the meta-system chooses an expert system (or expert systems) to solve a problem, the meta-system should be able to initialize this expert system (sub-process), constantly monitor its progress, and terminate this sub-process whenever necessary. This can be achieved through Lisp™ commands, such as FORK and WAIT. On the other hand, communication between the meta-system and expert systems can be made by allocating and maintaining a common area as one-directional pipes, which are monitored by the meta-system and controlled by the meta-knowledge.

Because all expert systems deal with highly interrelated problems, it is convenient for these expert systems to share the same working memory and conflict sets. It is easier to handle a homogeneous environment than a heterogeneous one, since the former has the same data representation and the same inference mechanism. The main tasks here are concurrence control and recency control. The concurrence control problems of homogeneous distributed expert systems resemble those encountered in a distributed database management system, and are crucial to the conflict-resolution strategy in OPS5™. Time tags are assigned by the meta-system to every expert system, and the meta-system decides the recency of each instant generated by different expert systems.

When dealing with heterogeneous distributed expert systems, e.g. when integrating into an integrated distributed intelligent system the expert systems that are implemented in other environments or tools (such as KEE™ or G2™), the main issue is how to map or translate different data representations and different inference mechanisms. The meta-system can possess the knowledge of its working expert systems, including the particular expert system's knowledge representation techniques, inference mechanisms, and control strategies. A better way may be to create an expert-systems-capability-knowledge-base, like the UNIX™ system that stores information about most of the terminals' capabilities in the database TERMCAP. It is only through this knowledge base that the meta-system can communicate with different expert systems and choose the right one to perform a certain task.

2.4.2 Implementation in C™ environment

Another meta-system is implemented with C™ language (Cha *et al*, 1991). There are four main reasons to use C™ as an implementation language:

First of all, C™ language is versatile in both numerical computation and symbolic manipulation. Its capability to handle numerical operation is much more powerful than Lisp™, Prolog™, OPS5™ and other expert system development tools. It is also superior to FORTRAN™ and BASIC™ in terms of symbolic processing operation. This advantage makes it easier to integrate different forms of knowledge with C™. Secondly, the C™ language possesses merits of both high-level and low-level languages such that it is very flexible and convenient for program coding and control on hardware, especially on the UNIX™ operating system that is developed in C™. Thirdly, the C™ language can easily access other language environments by interfaces written in mixture of C™ and assembly language. Finally, C⁺⁺™ is an object-oriented programming language, and is an extension of C™.

The advantage of implementing such a meta-system is that the symbolic process can follow the progress of the numerical process by receiving posted information from the numerical procedure at several steps during number-crunching execution. The symbolic process then has the option to continue the numerical procedure, to change some parameters, or to abort the procedure all together. This contracts favorably with shallow-coupled processes in which the heuristic process invokes a numerical routine via a procedure call, supplies the necessary input information, and passively waits for the numerical process to finish execution and provide the required output. Such a system configuration is demonstrated in Figure 2.5.

This new meta-system, namely Meta-C, is an intelligent system building tool that facilitates deep-coupled integration reasoning and algorithm-based numerical processes, which takes C™ as the fundamental knowledge representation language and serves as a functional supplement to C™ language. The important features of Meta-C are:

(1) flexible reasoning;

(2) a high-level representation language, which incorporates extended C™ data structures to define knowledge bases explicitly; and

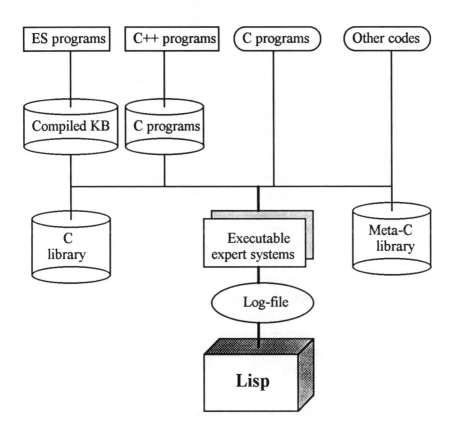

Figure 2.5 Information processing schematics of meta-C.

(3) the collaboration of different types of programs by supplying a flexible interface based on a news posting and delivering model.

Since Meta-C knowledge sources are compiled into standard C™ data structures, it has reasoning and knowledge representation facilities comparable to those of the existing tools, maintaining the real-time performance offered by low-level C™. Moreover, to ease data communication between a symbol manipulating heuristic process and a data-crunching numerical algorithm, the working memory is unified into a normal C™ data structure that is partitioned.

A knowledge source defines the scope of the partitioned working memory for it to access, defines the triggering preconditions for it to fire, defines the actions for it to take, and defines the initializing attributes before the knowledge source is used as well as the working team it belongs to, and so forth. Several related knowledge sources can be grouped together and a 'team head', the planner, stipulates how the knowledge sources interact and in what order.

The news posting and delivering model springs from the blackboard control architecture (Hayes-Routh, 1985) and acts as the mediator between heuristic reasoning and algorithmic numerical procedures, and thus serves as the synchronization mechanism between the two processes. Rather than allowing each knowledge source to search the blackboard for the information it requires, the strategy followed is to pass the newly received information through a fast pattern matching scheme and subsequently alter the knowledge sources that may have use for such information. This lowers the execution time and favors our goal of real-time performance. The pattern matching network is constructed by means of a compiler, after the knowledge sources are developed.

2.4.3 Implementation based on OOP technique

Based on the two implementations above, a more powerful meta-system has been developed using the object-oriented programming technique. The new meta-system Meta-COOP is coded in C⁺⁺, and run under UNIX™, VMS™ and DOS™ operating systems for SUN workstation, VAX and PC 486 machines. Meta-COOP provides such distinct characteristics as the integration of various knowledge representations and inference methods.

The process to solve a complicated manufacturing problem is actually a synthesizing process, which employs various knowledges and problem-solving strategies. In Meta-COOP, the organization structure of the knowledge-base can be divided into several components. For instance, the knowledge organization of a missile system can be separated into several subsystems as shown in Figure 2.6. Meta-COOP adopts the object-oriented programming technique and frame-based knowledge representation to implement the organization, management, maintenance and applications as a complex knowledge base system.

Meta-COOP distributes its meta-knowledge into many knowledge bases. Each knowledge base can be viewed as a fundamental *knowledge unit* (this may be a set of rules, a set of operation commands, or numerical models) to deal with a manufacturing problem and attached to a *frame* or an *object* that represents the problem. *Frame-based* or *object-based* knowledge representation not only describes in detail the attributes of an object but also hierarchically constructs the general knowledge base system, thus expressing the internal relationship between the knowledge bases. Such a hierarchy often has the two structures: classification structure and decomposition structure.

The decomposition structure stands for the organizational characteristics of an object in nature. Each part to be decomposed is always an organic part of the whole object. For instance, a missile system (Figure 2.6) can be decomposed into the basic missile, ground equipment and supplies facility; while the basic missile can be further divided into airframe, rocket engine, propellant systems, power supplies, armament and fusing, flight control system, and guidance system. If some parts in the decomposition structure are ignored, the whole system may lose its original characteristics. Such a decomposition structure can hierarchically represent the organizational construction and interrelations of knowledge bases.

The procedure to solve a complicated manufacturing problem consists of two stages: the first stage is to decompose a complicated problem into many basic sub-problems (from general to particular or from top to bottom). At the second stage, the sub-problems are solved individually. Meta-COOP provides the intelligent facilities for implementing this procedure. It resolves a complicated problem using a frame-based hierarchical structure and accomplishes communication and conflicting coordination between sub-problems through *message sending*. Each sub-problem or subsystem (it is always viewed as an object in Meta-COOP) and the interrelations can be described by the frame-based structure, while a knowledge unit to handle the sub-problem is closely attached to a *slot* in a frame. A knowledge unit may be constructed by a set of production rules, a set of operation commands, analysis methods, external procedures written in any other languages, and internal subroutines written in C™.

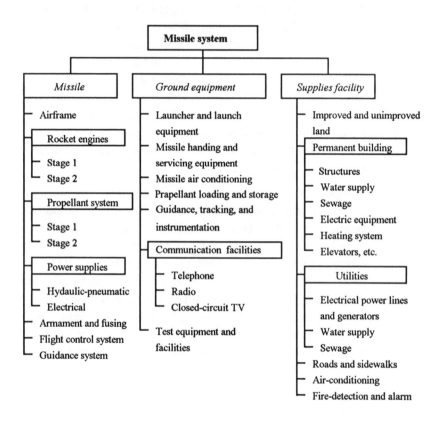

Figure 2.6 Hierarchical architecture of a missile system.

The classification structure describes the characteristics selection of a subsystem or sub-problem. For example, a missile engine can be classified as rocket engines or other types, while rocket engines may be grouped into solid and liquid rocket engines, and so on (see Figure 2.7). The classification relationship can be expressed by a *Superclass-Class-Subclass* hierarchical structure in Meta-COOP.

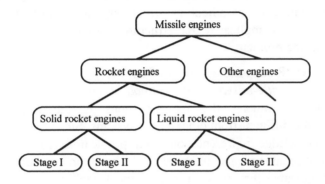

Figure 2.7 Classification structure of missile engines.

A missile engine is described by a frame that represents all common attributes and applies the knowledge about the missile engine. The frame is a Superclass frame. A Class frame here is the description of rocket engines, which includes the specific attributes and knowledge about rocket engines. The Class frame can inherit all common attributes and knowledge from its Superclass frame. Similarly, stage I solid rocket engine can be viewed as a Subclass frame that represents the specific features of the stage I rocket engine and inherits information from its Superclass and Class frames. In Meta-COOP, membership of the same classification, such as solid rocket engines or liquid rocket engines, will be directly expressed by *Memberof slot* in a frame.

It is also very convenient to extend the frame system, e.g. to add a new frame into the knowledge base. For instance, if stage III solid rocket engine in a catalog of solid rocket engine needs to be added, users only need to set up the frame that describes the specific attributes and knowledge about the stage III solid rocket engine, then connect this new frame with other frames by the Superclass–Class–Subclass relationship linkages.

With OOP technique, users can effectively design a hierarchical knowledge structure that makes the knowledge base systematized, modularized, and easier to understand, maintain and manage. Using the decomposition and classification structures, the meta-system can be expressed as a complicated tree structure available to represent a real manufacturing problem.

Meta-COOP supports a variety of knowledge representations such as frame-based, *rule-based* and *method-based* representations. The communication among the representations is the most important characteristic different from other expert systems. Since Meta-COOP uses various knowledge representations and problem-solving strategies, its inference mechanisms and information exchange in problem-solving are more complicated than those in conventional intelligent systems. Figure 2.8 demonstrates the fundamental control structure.

(1) *Method base*: A method base is a specific procedural knowledge that is attached to the frame. It consists of two parts: an *information filter* (a group of keywords) and a *method body*. The information filter determines if a method can be triggered or called when the *sending message* goes through the filter. The method body can perform symbolic reasoning and numerical computation, control the problem-solving process, and send a message to other frames.

(2) *Frame-rule base*: A frame is a fundamental unit in the meta-knowledge base. It expresses the natural attributes of an object with *slots* and *facets*. It also clearly represents the topological relationship between objects with the decomposition and classification structures. A set of rules as a *slot value* is stored in a slot of the current frame. This slot is named as the *rule slot*.

(3) *Database*: The meta-system has a miniature relational database to record the intermediate results.

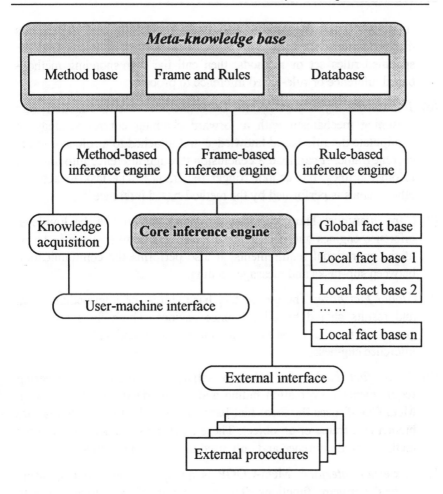

Figure 2.8 Integration of inference strategies in Meta-COOP.

(4) *Method-based inference*: The inference engine is used to perform a method body, to implement numerical computation, and to select search paths. In addition, it can also invoke complex external procedures.

(5) *Frame-based inference*: This performs an evaluation manipulation on the current frame, and inherits information from other frames. The

frame-based inference accomplishes data processing, and implements the evaluation on a method or a set of rules, that is, it can find the specified rules set or methods, then call the corresponding method-based inference or rule-based inference to process.

(6) *Rule-based inference*: Meta-COOP includes the production rule reasoning mechanism with a forward chaining control strategy to manipulate the rule-based knowledge unit attached to a rule slot in the current frame. It can load a conclusion through inference into any slot in other frames, but cannot trigger the frames. The task to activate other frames is performed by the method-based inference.

(7) *Core inference*: The core inference engine can invoke the low-level inference engines and coordinate the conflicts between them in terms of the operational requirements. It also performs the communication between sub-tasks and message sending.

(8) *Global fact base*: The global fact base stores the intermediate facts and results shared by all inference engines. These facts are often generated by the rule-based inference engine and called by other inference engines.

(9) *Local fact bases*: In order to satisfy the software engineering requirements (information hiding and modularized knowledge base), Meta-COOP uses the local fact bases to store the local facts that are hidden in a frame or an object. These local facts are derived from the method-based inference and only used in evaluating methods.

(10) *External interface*: Meta-COOP is an open system to help users extend software functions. The external interface allows users to define some functions or procedures written in C as the internal functions of Meta-COOP. The functions are called by the method-based inference engine. The external interface also sets up the communication facility for message sending and result returning.

(11) *Knowledge acquisition*: The module facilitates editing and compiling a knowledge base. The source codes of the knowledge base can be generated by common editors such as VI™, PE2™, etc., then stored as ASCII files. After the files are compiled, the ASCII file is transferred into the binary file that is directly referred to the inference

engines.

(12) *Man–machine interface*: This receives the input information from users and provides the problem-solving results and interpretation of reasoning processes. Meta-COOP interface is written by SunView™ and controlled through a mouse as shown in Figure 2.9.

[FILE]: managing the knowledge base files and data files

[REQUIRE]: receiving the input information from uses

[LOAD]: loading the compiled knowledge base files

[COMPILE]: compiling the specified project file

[EDIT]: editing a knowledge base file written in ASCII

[PROJECT]: displaying the content in a project file

[SYS_STRCT]: displaying the Meta-COOP software configuration and knowledge base structure by graphics

[SIMULATE]: calling external simulation programs

[LAYOUT]: calling computer graphics packages

[RESULT]: displaying final results

[HELP]: explaining the operational commands and procedures

[ANALYSIS]: calling external numerical computation programs

In the next chapter, we will discuss the key issues and methodology for implementing Meta-COOP, including object-oriented programming (OOP) techniques, knowledge representations, operation principles of various inference engines such as method-based, rule-based, frame-based and core inference engines, as well as the interpretation strategies of Meta-COOP.

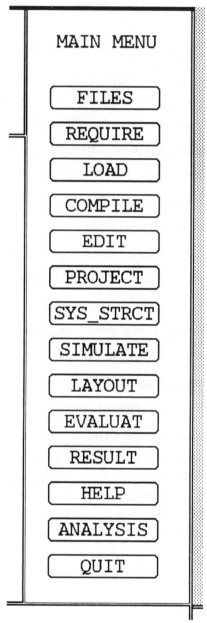

Figure 2.9 Demonstration of Meta-COOP interface.

3

Meta-COOP implementation

3.1 Object-oriented programming technique

To successfully implement the functions of the meta-system, a new structure of the meta-system, namely Meta-COOP, has been developed based on the object-oriented programming (OOP) technique. This chapter first introduces some basic concepts about OOP techniques and compares it with other conventional programming languages, then discusses the knowledge representations and knowledge base organization structure, and presents four inference functions and algorithms; i.e., *frame-based inference, method-based inference, rule-based inference* and *core inference*. Practical examples are described. Finally, the interpretation strategy of Meta-COOP is addressed and a case study is presented.

Object-oriented programming (OOP), like other information-oriented software design methodologies (such as data-flow-oriented design and data-structure-oriented design), provides a powerful representation for a real-world problem and maps it into a solution domain in software. Unlike other software design methods, OOP results in a design to interconnect data objects and processing operations in such a way that a software (or a program) modularizes information and processing rather than processing alone (Pressman, 1987). In conventional software design, data and processing operations are fully separated.

OOP implements three important software design concepts, that is, *abstraction, information hiding,* and *modularity*. Information hiding is the most important feature of OOP among existing software programming methods. Only OOP enables software designers to implement the three concepts without complexity or compromise. Wiener and Sincovec (1984) brilliantly summarized OOP methodology as follows:

No longer is it necessary for the system designer to map the problem domain into pre-defined data and control structures present in the implementation language. Instead, the designer may create his or her own abstract data types and functional abstractions and map the real-world domain into these programmer-created abstractions. This mapping, incidentally, may be more natural because of the virtually unlimited range of abstract types that can be invented by the designer. Furthermore, software design becomes de-coupled from the representational details of the data objects used in the system. These representational details may be changed many times without any fallout effects being induced in the overall software system.

Objects and *operations* are not new programming concepts, however object-oriented programming is. At the very early development stage of numerical computation, assembly languages enabled programmers to use machine instructions (operators) to manipulate data items (operands). The level of abstraction that was applied to the solution domain was very low.

As high-level programming languages (e.g., FORTRAN™, ALGOL™ and COBOL™, and so on) appeared, objects and operations in the real-world problem space could be modeled by predefined data and control structures that were available as part of high-level languages. Software design focused on the representation of procedural details using a selected programming language.

During the 1970s, concepts such as data abstraction and information hiding were introduced, and data-driven design methods emerged, but software designers still concentrated on process and its representation. At the same time, modern high-level languages (such as Pascal™) introduced a much richer variety of data structures and types.

While conventional high-level languages were evolving during the late 1960s and the 1970s, a new class of simulating and prototyping languages such as SIMULA™, was developed. In these languages, data abstraction was emphasized and real-world problems were represented by a set of objects to which a corresponding set of operations were attached. The use of these languages was radically different from the use of more conventional languages.

The OOP methodology has evolved during the past 15 years. The early work introduced the concepts of abstraction, information hiding, and modularity into software design to enhance program quality. During the 1980s, the first OOP language Smalltalk80™ was released, which stimulated the development and applications of OOP technology. Today, OOP is widely used in software design and applications that range from computer graphics (Lorensen, 1986b), telecommunications (Love, 1985), manufacturing control systems and shop floor control systems (Hinde *et al.*, 1992; Smith and Joshi 1992), CAM/CAM, as well as CIMS, and so on.

OOP has the following five important features:

- *structural object description* (modularity) and *hierarchical structure* based on *object classification*, and *information inheritance* in the same family;
- data (information) hiding and data abstraction;
- *process-operations* on data stored (or connected) in a hierarchical structure of objects;
- information communication performed by *message sending* and inheritance; and
- progressively detailed programming procedure.

The most obvious advantage of an object-oriented software system is that the system can store *static data* and *dynamic operations* about an object in a structural *frame*. Because of the characteristic that OOP requires to hide all data and abstractly represents data models, any external operation routine cannot process the internal data stored in an object. In other words, OOP structure can establish a *specific wall* so that the external operation procedure cannot process the internal data. A data-processing operation is connected with some data items and stored in a so-called *object frame*. OOP can integrally describe the attributes of an object (static information) and the special operations on these attributes (dynamic information). The two types of information may be viewed as an integral entity and organized as a special object frame. The communication between the object and external data items or external operations can be executed by the so-called message sending.

Superclass and *subclass* relationships among objects as well as some inheritance principles enable all subclass objects to partly or fully inherit the static data and dynamic operations stored in the superclass. In other words, a superclass object describes the common attributes that all its subclass objects have. As a result, the only consideration of a software designer, in setting up a special subclass object, is how to describe and operate those specialized attributes that the subclass object have. The common attributes and operations that all subclass objects possess have to be stored in a superclass object and can be shared by its all subclass objects when software running.

The knowledge about an object therefore can be easily modularized with such an object-oriented design method and frequently referred to by its subclass objects through some inheritance principles. It is very easy to build a new object because software designers only consider those data items, attributes and operations different from those of its superclass object.

In addition, because knowledge representation adopts the object-oriented structure, other partitions will never be changed when a part of the knowledge base is locally modified. Adding new data structure in the knowledge base does not need to modify the data-processing operations because of data abstraction. These advantages are very significant when developing and maintaining a large-scale knowledge base system.

3.2 Objects, operations and messages

Generally speaking, a software is functioned in such a way that a data structure (of varying levels of complexity) is acted upon by one or more processes according to a procedure defined by a static algorithm or dynamic commands. Thus, Meta-COOP design should provide a mechanism for: (i) representation of data structures, (ii) specification of operation processes, and (iii) strategy of references and inheritances.

From the viewpoint of OOP methodology (Lorensen, 1986), an object is a component of the real-world, which is mapped into the software domain. In programming, an object is typically a producer or consumer of information or an information item. For example, some typical objects might be machines, foods, fruit, commands, files, displays, switches, signals, strings, and so on. When an object is mapped into its software implementation, it consists of a *private data structure* and several *processes*, called operations or methods, which may legitimately transform the data structure. Operations contain control commands and procedural statements that may be invoked by other objects through message sending. The data structure implementation of an object is illustrated in Figure 3.1.

Referring to Figure 3.1, a real-world object (a mill machine) is mapped into a software implementation. The figure exhibits a private data structure and related *operations*. The data structure may take the form depicted in Figure 3.2. Entries in a mill machine are composed of its structural descriptions, attributes, as well as attribute values. A picture or diagram may also be contained within an entry. The object, mill machine, also contains a set of operations (e.g. startup, shutdown or machine a piecework) that can process data items in the data structure described above. The private part of an *object* is the data structure and the set of data-processing operations.

An object also has an *interface*. *Messages* move across the interface and specify what operation on the object is desired, rather than how the operation is to be performed. The object can determine how the operation requested is to be implemented when it receives a message. Whether or not the set of operations of an object is activated by a message sending depends on the matching *keywords*. Only when a predefined interface fully matches the messages received, can the operations of the object be carried out and a new message sent to another object.

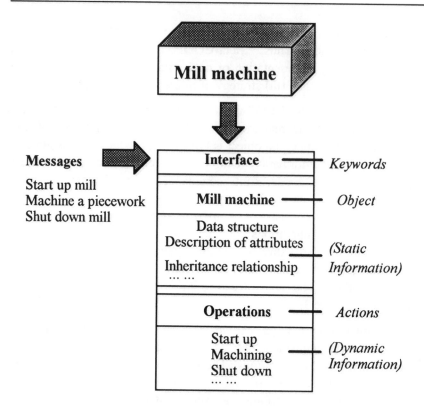

Software implementation of object

Figure 3.1 Objects, operations and messages of Meta-COOP.

By defining an object with a private part and providing messages to invoke appropriate processing, information hiding is achieved; that is, the details of implementation are hidden from all program elements outside the object. Objects and their operations provide inherent modularity, i.e. software elements (data and process) are grouped together with a well-defined interface mechanism.

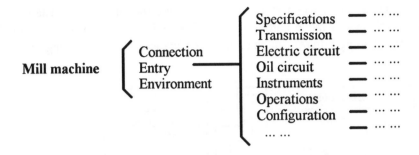

Figure 3.2 Data structure of a mill machine.

3.3 Classes and instances

Many objects in the real-world have similar *attributes* and perform similar operations. For instance, there are mill machines, drill presses, and jig borers in a manufacturing plant. Although these objects are different from each other, all belong to a larger *class* called the metal cutting tool. All objects in the metal cutting tool class have attributes in common (e.g. all use electric motors) and perform common operations (e.g. all perform metal-cutting). Therefore, by categorizing a mill machine as a member in the metal cutting tool class, its attributes and operations can be estimated even without the details about its functions.

Software implementations of object in the real-world are categorized in the same way. All objects are the members of a larger class (superclass) and inherit the private data structure and operations that have been defined for that superclass. In other words, a superclass is a set of objects having the same characteristics. An individual object is therefore an *instance* of this superclass.

As an example of the relationship between objects and classes, the object mill machine is an instance of class metal cutting tool and inherits the data structure and operations of a metal cutting tool. Therefore, the metal cutting tool becomes a data abstraction that enables software designers to define each instance in a facile manner. For example, 'type mill machine IS INSTANCE OF metal cutting tool' implies inheritance of the attributes of metal cutting tool. In Figure 3.3, other objects (e.g. drill presses and jig borers) can be defined as instances of metal cutting tool.

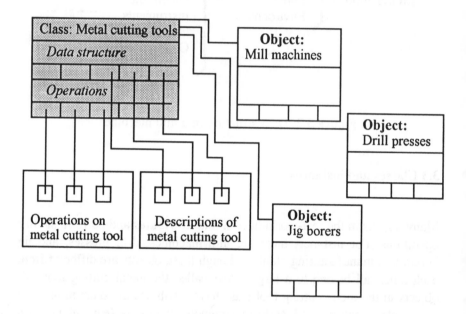

Figure 3.3 Objects, classes, and inheritance of Meta-COOP.

The use of classes and inheritance is crucially important in modern software engineering. Reusability of program components is achieved by creating objects (instances) that are built on the existing attributes and operations inherited from a class. Software designers only need to specify how a new object differs from the class, rather than having to define all characteristics of the new object.

Unlike other design concepts that are programming-language-independent, the implementation of classes and objects varies with the programming language to be used. For this reason, the preceding generic discussion may require modification in the context of a specific programming language. For example, Ada™ implements an object as a package and achieves instances through using data abstractions and typing. Smalltalk80™ (an OOP language), on the other hand, implements each of the above concepts directly.

3.4 Methods and information sending

OOP employs a class structure as an object representation or description. In Meta-COOP, a *unit* structure (or frame structure) is used to represent classes and instances (or objects) as well as their hierarchical relationships or inheritance relationships. A unit (also called frame) is similar to a *record* in database, but it is different from record since a unit includes not only the static descriptions about an object but also its dynamic operations stored in a *method slot*. A unit often contains several slots that can represent the attributes of an object and the inheritance relationship between objects, and store various operations and commands.

A unit is a powerful data structure to represent objects in the real world. Besides the common characteristics that conventional languages have, such as modularized data structure, a unit has a special slot, called method, which is usually used to represent dynamic knowledge existing in the real world; i.e. processing operations on data and information.

Operations or commands stored in a method slot can be triggered by a message sending process, and the commands in the method slot can send a message to other method slots. The message sending process is performed by a core inference engine in Meta-COOP. The core inference engine has to know all the information about which unit a message will be sent to and which method in this unit will be performed. However, when the sending information form matches the needed information form in the method slot, the core inference engine can perform a procedure stored in the method slot. In Meta-COOP, operations and commands in a method slot can be represented by such programming language as Pascal™ (the compiler of method slot is specifically written only for Meta-COOP). This simplified language can be viewed as a subset of standard Pascal™, and complied by the specialized compiler.

3.5 Active value and facets

In Meta-COOP, when a part of the knowledge base (or an object) is modified, the action automatically causes changes of the other parts or conducts some operations on other objects. When solving real, complicated engineering problems, the unit's operations and knowledge representations are very important. The automatic features can be obtained using *active value* that looks like a 'firer' whenever and wherever it might emerge. In a unit, each slot has various *facets* that represent and limit the range of *slot value*. Meta-COOP uses active value facet to represent a set of functions or commands.

3.6 Inheritance

Based on OOP method, an intelligent system development environment should be able to set up and organize knowledge models efficiently. In Meta-COOP, all units are organized in a hierarchical architecture of knowledge representation. Meta-COOP environment provides such *links* as *member*, subclass and superclass, as well as a set of prototypical frames to express frame information. The prototypical frames here are called class frames. Every frame (unit) can be connected with one or several class frames through the member slot. These class frames can inherit all information and data from their superclass frame connected by a member slot. The information to be inherited is often stored in a member slot that includes facets and values.

Meta-COOP provides a powerful mechanism to implement such an inheritance operation. A subclass can inherit all information and data stored in its superclass. The principles that inheritance must follow are determined by the content in an inheritance slot. Currently, inheritance operations include only *union* and *override* operations.

Inheritance operations play a very important role in the development and maintenance of meta-knowledge base. The hierarchical structure formed by class–subclass–member can be used to organize a set of interrelated objects and to implement the modularized knowledge management efficiently. In Meta-COOP, all *rules* are divided into basic rule groups according to application domain and specifications. The small rule groups, viewed as a heuristic-*knowledge element*, may be embedded into a special *rule slot* and performed when solving a special domain problem. Dividing a large rulebase into small heuristic-knowledge elements according to the problems to be solved can enhance the efficiency and quality of the organization, management and maintenance of the meta-knowledge base.

3.7 Rules

Rules used in Meta-COOP are production rules, which can represent dynamic behaviors of objects and expert's heuristic knowledge. Rules may be divided into many small sets of rules according to real problems to be solved and domain knowledge. A set of rules that is often viewed as an integrated *knowledge element* is embedded in a rule slot of a frame and becomes the value of the rule slot. To use the set of rules, a rule-based inference engine is invoked when the rule slot and the inference engine type are specified. In addition, by the inheritance principle, a set of rules in superclasses can be invoked by the *current frame,* i.e. the current frame can inherit all rules sets in its superclass to solve an engineering problem that is defined in the current frame.

With such a knowledge representation structure, more subclasses can share the knowledge stored in their superclasses; that is, the knowledge base in a superclass may be sequentially referred by its subclasses. When setting up an object, since common attributes of the object have been described in its superclass, only the special attributes and operations (including static and dynamic) of the object are classified and all common attributes and operations are inherited. As a result, the meta-knowledge base is much easier to develop and maintain.

3.8 Frame-based knowledge representation and inference

The frame representation is very often used in intelligent systems. A characteristic of its data structure is to efficiently describe physical objects and their relationship in detail. With numerous slots, a frame system forms a hierarchical configuration of knowledge base. In Meta-COOP, the syntax of a frame representation is formalized as follows:

$<$Frame$>$ :: = unit $<$Frame name$>$ in-knowledge-base $<$Knowledge base$>$

$\quad$ {Superclasses: {Ancestral frame name}$^+$}*
$\quad$ {Subclasses: {Descendant frame name}$^+$}*
$\quad$ {Member: {Membership frame name}$^+$}*
$\quad$ {Memberof: {Type frame name}$^+$}*
$\quad$ {Slot}$^+$

$<$Slot$>$ :: = Memberslot | Ownslot: $<$Slot name$>$ from $<$Frame name$>$

$\quad$ Valueclass: $<$Slot type$>$
$\quad$ {Inheritance: $<$Inheritance type$>$}*
$\quad$ {Coordinate.Min: $<$Integer$>$}*
$\quad$ {Coordinate.Max: $<$Integer$>$}*
$\quad$ {$<$User definition facet$>$ $<$Facet values$>$}*
$\quad$ {Value: $<$Values$>$}$^+$

$<$Slot type$>$ :: = Integer | Real | String | Rules | Methods | $<$Frame name$>$
$<$Inheritance type$>$ :: = Override.Value | Union | Methods
$<$Value$>$ :: = Integer | Real | String | $<$Rules name$>$ | $<$Frame name$>$ |
$\quad\quad\quad\quad$ $<$Method name$>$

where

$\quad$ {}* expresses optional,
$\quad$ {}$^+$ expresses an occurrence at least, and
$\quad$ | expresses "or".

Override.value stands for the override inheritance principle of slots. The principle definition is that a value of the current frame slot can be directly embedded in the current frame slot if the value exists. Otherwise, the slot has to inherit a value (the value may be an attribute value, an operation or a set of rules) from its ancestral frame that is defined by superclass and memberof slots. The ancestral frames include all generation-to-generation ancestral frames.

Union expresses the union inheritance principle. By definition, a union value, through coupling the current value with the inherent value from the ancestral frames, as a new slot value, is loaded in the current frame slot.

Method is a kind of special slot and represents the inheritance principle of method. The principle allows a frame to inherit all operations stored in a method of its ancestral frame. Information sending defines which method in a frame is triggered.

Valueclass slot is used to record the types of a slot value. If the slot loads a method that includes a set operations, valueclass records the name of the method; if the slot stores a set of rules, valueclass writes down the first address that is specified after a set of rules (knowledge elements) is compiled. These rules are only used by the rule-based inference engine, rather than by other operations.

Coordinate.Max and Coordinate.Min slots define the range of the value in valueclass if the value is numerical information. Coordinate.Max slot determines the upper limit of a numerical value, while Coordinate.Min slot gives the lower limit.

The main task of frame-based inference is to implement the operation on slot values of a frame. The operational realization depends on the hierarchical structure of a frame system, inheritance characteristics, as well as the hidden relationship between ancestral and descendant frames.

The frame-based inference engine first checks valueclass restraints and coordinate restraints, then transforms various data types. Finally, it carries out operations such as adding a slot value or deleting one.

The following is a frame-based inference algorithm used in Meta-COOP. Its program control flow diagram is shown in Figure 3.4.

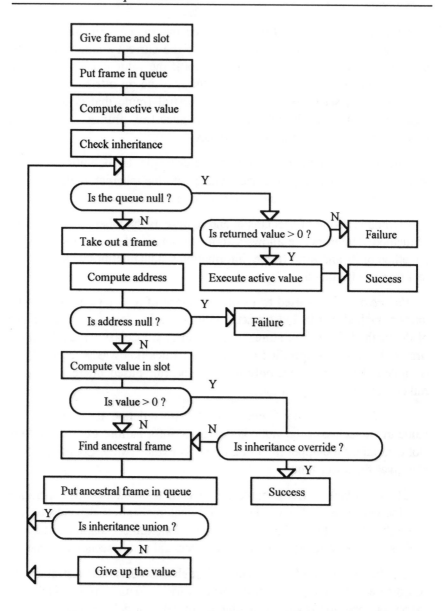

Figure 3.4 Program control flow diagram of frame-based inference.

Algorithm of frame-based inference:

(1) Give frame name and slot name.

(2) Put the frame name into the queue.

(3) Compute the activevalue in a slot.

(4) Check the inheritance characteristic in inheritance slot.

(5) Check whether the queue is empty. If it is null and the returned value is larger than zero, then exit with success. Otherwise display a failure message and exit. If the queue is not null, continue.

(6) Take out a frame name from the queue.

(7) Compute the slot address of the frame.

(8) Check whether the address is null. If the address is empty, display a failure message and exit. Otherwise, continue.

(9) Compute the value in the slot.

(10) Check whether the value is less than zero. If negative value and the inheritance characteristic is override, exit with success. If positive value, continue.

(11) Find all ancestral frames of the current frame.

(12) Put the ancestral frames in the queue.

(13) Check whether the inheritance characteristic is union. If yes, go to (5). Otherwise, continue.

(14) Give up all values and go to (5).

In the frame-based inference algorithm, the reasoning results could be either common data types (such as string, real and integer) or an address of a set of rules, as well as an address of a method.

3.9 Rule-based knowledge representation and inference

Production rule is an important kind of knowledge representation and is widely used in intelligent systems. In Meta-COOP, production rules are divided into many small sets of rules (e.g. small *heuristic-knowledge elements*) based on a special application domain and attached to the rule slot of a frame. In terms of efficiency, the reasoning process is much faster because a rule-based inference engine deals only with a small set of rules. On the other hand, all heuristic-knowledge elements attached to a frame can be invoked by its subclasses through the inheritance relationships. The following is the formalized representation of a rule in Meta-COOP.

$<$Rule$>$:: = (fact $\{<$Premise$>\}^+$ then $\{<$Conclusion$>\}^+$)
$<$Premise$>$:: = ($\{<$Function$>^*\}$ $\{<$Element$\}^+$)
$<$Function$>$:: = and, or, not, +, -, ... , any predicate or operator
$<$Element$>$:: = $<$Vector element$>$ | $<$Attribute value$>$
$<$Value$>$:: = $<$Symbol$>$ | $<$Number$>$
$<$Conclusion$>$:: = ($\{<$Statement$>\}^+$ | $\{<$Action$>\}^+$)
$<$Statement$>$:: = ($<$Vector element$>$) | $<$Attribute value$>$)
$<$Action$>$:: = $\{<$Function$>\}^+$ | $\{<$Operator$>\}^+$

where

$\{\}^*$ means optional,
$\{\}^+$ means an occurrence at least, and
| means "or".

As an example, the knowledge base for the conceptual design of a missile weapon system is discussed, which is implemented by Meta-COOP. With hundreds of rules and frames, this system can provide the preliminary conceptual design schemes to designers. The following is an example of a heuristic-knowledge element (or a set of rules).

Memberslot: S_engine_rules from power_sys
    Inheritance: Override.Values
    Valueclass: RULES
    Coordinate.Min: 1
    Coordinate.Max: 1

Values: {
    rule 531

        fact T_requirements.Range>=100
        and T_requirements.Flight_velocity>=1.2
        then _FRAME.M_Engine_type="Ramjet_engine";
           Engine_type="Ramjet_engine"

    rule 532

        fact T_requirements.Range>100
        and T_requirements.Flight_velocity<=0.9
        then _FRAME.M_Engine_type="Turbojet_engine";
           Engine_type="Turbojet_engine"

    rule 533

        fact T_requirements.Range<=100
        and T_requirements.Range>70
        and T_requirements.Flight_velocity>=2.0
        then _FRAME.M_Engine_tyjet_engine";
           Engine_type="Ramjet_engine"

    rule 534

        fact T_requirements.Range<=70
        and T_requirements.Flight_velocity<=0.9
        then _FRAME.M_engine_type="Solid_rocket_engine";
           Engine_type="Solid_Rocket_engine"

    rule 535

        fact T_requirements.Range>70
        and T_requirements.Range<=100
        and T_requirements.Flight_velocity<=0.9
        then _FRAME.M_Engine_type="Liquid_rocket_engine";
    Engine_type="Liquid_Rocket_engine"

    rule 536

        fact Engine_type="Turbojet_engine"
        and Turbojet_type="FW-41"
        then T_requirements.Max_diameter=0.76

    rule 537

        fact Engine_type="Turbojet_engine"
        and Turbojet_type="TRI60-3"
        then T_requirements.Max_diameter=0.54

    rule 538

        fact Engine_type="Solid_rocket_engine"
        and T_requirements.Range<=70
        then T_requirements.Max_Diameter=0.36

    rule 539

        fact Engine_type="Solid_rocket_engine"
        and T_requirements.Range>70
        and T_requirements.Range<=100
        then T_requirements.Max_Diameter=0.54

```
rule 540
        fact Engine_type="Solid_rocket_engine"
        and T_requirements.Range>100
        then T_requirements.Max_Diameter=0.76
rule 541
        fact Engine_type="Liquid_rocket_engine"
        and T_requirements.Range<=70
        then T_requirements.Max_Diameter=0.36
rule 542
        fact Engine_type="Liquid_rocket_engine"
        and T_requinge>70
        and T_requirements.Range<=100
        then T_requirements.Max_Diameter=0.54
rule 543
        fact Engine_type="Liquid_rocket_engine"
        and T_requirements.Range>100
        then T_requirements.Max_Diameter=0.76
rule 544
        fact T_requirements.Missile_type="ground_ship"
        then boosterdesign="Yes"
rule 545
        fact T_requirements.Missile_type="ground_air"
        then boosterdesign="Yes"
rule 546
        fact T_requirements.Missile_type="ground_ground"
        then boosterdesign="Yes"
rule 547
        fact T_requirements.Missile_type="ship_ground"
        then boosterdesign="Yes"
rule 548
        fact T_requirements.Missile_type="ship_ship"
        then boosterdesign="Yes"
rule 549
        fact T_requirements.Missile_type="ship_air"
        then boosterdesign="Yes"
        }

Memberslot: design_method from power_sys
  Inheritance: METHOD
  Valueclass: METHODS
  Values: power_design_method

METHOD power_design_method (design: keyword)
VAR
  design: keyword;
```

```
   M,L: real;
BEGIN
   reason (_FRAME, "S_engine_rules");
   def_member ("Engine",Engine_type);
   if (boosterdesign = "Yes") then
          BEGIN
                    send (design) to "Booster_p";
          END;
   if (boosterdesign = "No") then
          BEGIN
                    Mb:=0;
                    Lcb:=0;
          END;
   send (design) to "Engine";
   M: =Mb+Ms;
   _FRAME.Power_total_mass:=M;
   L:=Lcb+Lcs;
   _FRAME.Power_total_length:=L;
END.
```

In this example, the heuristic-knowledge element is attached to the 'S_engine_rules' frame that comes from superclass power_sys; i.e. the frame can inherit the knowledge from the superclass frame. The rules are used to choose types of rocket engines and size parameters. In addition, the example also includes a method slot (e.g. design_method) that triggers the rules.

The method has an activevalue that records the name of method (power_design_method) and the *method body*. When activevalue receives a message from other frames and successfully matches it with keywords in the method, the method body is activated and all operational commands in it will be executed. In the method body, command 'reason (_FRAME, "S_engine_rules")' means to choose a rocket engine type using the rules stored in S_engine_rules frame. Command 'send (design) to "Booster_p"' will trigger the Booster_p frame if designing the booster of a missile is needed. Similarly, command 'send (design) to "Engine"' will activate the engine frame for determining the other parameters of a rocket engine.

Meta-COOP employs the control strategy of *forward chaining reasoning* (data-driven) to manipulate rules. The rule-based inference engine is very popular in commercial expert system tools, therefore, we will not discuss the algorithm of rule-based inference in detail.

3.10 Method-based representation and inference

A method slot can be used to express the procedural knowledge or a set of operations that are part of dynamic knowledge in Meta-COOP (the other part of dynamic knowledge is production rules). A method may be triggered by a message sent in the process of information transmission.

In Meta-COOP, the syntax of method can be expressed as:

\<Method\> :: = METHOD \<Method name\> \<Information table\>
            \<Definition of local variables\> \<Method body\>

\<Method name\> :: = \<String\>

\<Information table\> :: = ({\<Variable name\> : \<Variable types\>}*)

\<Definition of local variable\> :: = VAR {\<Variable name\>
            \<Variable type\>}*

\<Method body\> :: = BEGIN {\<Statements\>}* END.

\<Statements\> :: = \<Assignment statements\> | \<Conditional statements\> |
            \<Sending message statement\>

\<Assignment statements\> :: = \<Variable name\>: ={\<Variable name\> |
            \<Functions\>}$^+$

\<Functions\> :: = +, -, /, sin, cos, tan, log, exp, etc.

\<Conditional statements\> :: = If \<Conditions\> then BEGIN \<Conclusion\>
            END;

\<Conditions\> :: = {\<Variable name\> \<Operators\> \<Variable name\>}$^+$

\<Operators\> :: = \<, \>, \<\>, \<=, \>=, NOT, AND, OR, etc.

\<Conclusions\> :: = ({\<Statements\> | \<String\>}$^+$)

\<Sending message statement\> :: = send \<Information table\> to \<Frame
            name\>

\<Information table\> :: = ({Variable name}$^+$)

\<Variable name\> :: \<Frame name\> :: = \<String\>

<Variable types> :: = Keyword {<String> | <Integer> | <Real>}$^+$

Where

{}* expresses optional,

{}$^+$ expresses an occurrence at least, and

| expresses "or".

The method slot can perform message sending, control reasoning paths, trigger rules, carry out the complicated numerical computation, as well as call external procedures and data files. The following example is a complicated computing process to determine some parameters of a solid rocket engine. In the method body, statements 'Cya: = chz (alpha, 4, "qdxs.dat")' and 'alpha: = qdjs (qs, G1, 4, "qdxs.dat")' are two examples for referring external data files.

```
Memberslot: C_engine_parameter from Solid_rocket_engine
    Inheritance: METHOD
    Valueclass: METHODS
    Coordinate.Min: 1
    Coordinate.Max: 1
    Values: C_s_rocket_parameter

METHOD C_s_rocket_parameter (design: keyword)
VAR
    t, v, S, Cx, Q, pvs, G1: real;
    Fave, Mps, qs: real;
    Mpb, lamda_s, e, g, p, rp, nrs, Lps, Lh, D, Isigma2: real;
BEGIN
    reason (_FRAME,"C_engine_rules");
    t:=T_requirements.Range*1000/(T_requirements.Flight_velocity*340);
    _RRAME.Action_time:=t;
    g:=9.8;
    p:=1.25;
    v:=T_requirements.Flight_velocity*340;
    _FRAME.Engine_Diameter:=T_requirements.Max_diameter;
    D:=_FRAME.Engine_Diameter;
    M0:=T_requirements.Missile_mass;
    s:=0.25*3.1415926*D**2;
    G1:=(M0-Mb)*g;
    qs:=1/2*p*v*v*s;
    alpha:=qdjs(qs,G1,4,"qdxs.dat");
    _FRAME.Equil_attack_angle:=alpha;
```

```
Cya:=chz(alpha,4,"qdxs.dat");
pvs:=qs*Cya;
Cx:=chzx(alpha,4,"qdxs.dat");
Q:=1/2*p*v*v*s*Cx;
Fave:=Q;
_FRAME.Engine_thrust:=Fave;
Mps:=Fave*t/Is2;
Mps:=Mps*(1+0.03);
Isigma2:=Fave*t;
_FRAME.mass_of_fuel:=Mps;
_FRAME.Total_impulse:=Isigma2;
lambda_s:=0.64;
delta_lambda_s:=0.06;
lamda_s:=lambda_s*(1+delta_lambda_s);
Ms:=Mps/lambda_s;
_FRAME.Total_mass_of_engine:=Ms;
rp:=1.7*1000;
nrs:=0.92;
Lps:=4*Mps/(3.1415926*D**2*rp*Nrs);
Lh:=2*1/4*D;
Lcs:=Lps+Lh;
_FRAME.Engine_Length:=Lcs;
END.
```

In the representation of a method, a special variable type keyword is used in Meta-COOP. The value of keyword is itself. The objective for setting up keyword is to build an exchange standard for information transmission. Keywords in the information table determine if a method can be triggered. When the sending keywords fully match the keywords predefined in a method, the method may be invoked. For example, an information table in a method is defined as follows:

METHOD Parameter_design (design: keyword x: real)

The method will be triggered by a message such as 'design 4.3' or 'design 1988' etc., because the parameter in 'design' matches the keyword in the information table, i.e. x matching 4.3 or 1988.

A method in Meta-COOP is a set of statements or a group of operations. It can perform numerical computation and drive an inference engine to carry out symbolic reasoning operations (i.e. call the frame-based inference engine to accomplish symbolic evaluation). Of course, the process that alternately invokes symbolic inference and numerical computation is accomplished by the core inference engine (see section 3.11).

Another important role of method in Meta-COOP is to control the overall integrated distributed intelligent system. In Meta-COOP, the process of information sending implements the general problem-solving strategy of the meta-system. In this case, rule-based inference is used only inside a special frame where a set of rules is stored for a specified domain problem to be solved.

The general control strategy is determined by the message sending statements in method bodies. The process of message transmission decides the direction of the general inference, that is a supervising reasoning control strategy. The meta-system uses the characteristic to implement problem-solving, communication, conflicting coordination, task planning and optimal decision-making.

Even though a method description of Meta-COOP could support production rules, the conditional statements (clauses) are always used to switch operations. The conditional statements can be embedded in each other. The following lines express the method-based inference algorithm. Figure 3.5 demonstrates its program control flow diagram.

Algorithm of the method-based inference:

(1) Give a starting pointer of a method.

(2) Set the definitions of local variables in a method.

(3) Set the pointer aiming at the first statement.

(4) Take out a statement.

(5) Check whether the statement is an assignment (or a conditional one). If yes, execute the logic or numerical computation (the process is carried out by a subroutine), and go to (9). Otherwise, continue.

(6) Determine whether the statement is a rule-based reasoning request. If yes, perform a symbolic reasoning operation, and go to (9). Otherwise, continue.

(7) Check whether the statement is a request for sending message. If yes, send the message, and go to (9). Otherwise, continue.

(8) Carry out functional operations.

(9) Set the pointer aiming at the next statement.

(10) Determine whether pointer is empty. If yes, finish the inference. Otherwise, go to (4).

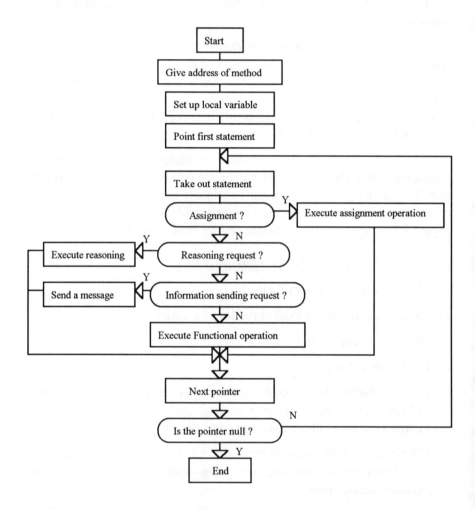

Figure 3.5 Program control flow diagram of method-based inference.

In the manipulation of method-based inference, since a method slot of a frame can send a message to trigger method bodies in another frame, the processing operation on a method slot is actually a recursive procedure (including direct and indirect recursion). The message sending process of method-based inference engine is a complex operation. First of all, the method-based inference engine requests the core inference engine to send a message to a method slot of a frame, then triggers the method body if the message can match to the keywords. Implementing the process needs to recursively invoke the method-based inference engine. Meta-COOP allows software designers to implement read/write files operations, design an interactive interface in a method body, define a variety of external functions (written by C™), carry out the interface definition, and invoke operation of external sub-programs with the method-based engine. As discussed previously, the definitions and applications of method-based inference engine have very important characteristics different from conventional symbolic inference engine. The use of OOP techniques in designing Meta-COOP greatly improves the capability and efficiency of IDIS in solving the real-world problems.

3.11 Core inference

The knowledge base in Meta-COOP provides a variety of knowledge representations and problem-solving strategies. Therefore, it is very important to make all inference engines (frame-based, method-based and rule-based inferences) work together to solve complicated engineering problems. In Meta-COOP, the process of message sending controls reasoning during problem solving. In order to coordinate various inference engines and the process of message sending, a higher-level inference engine, called the core inference engine, is required. The core inference engine carries out higher-level inference tasks (for example, sending a message to trigger a set of rules or a mathematical model), responds to requests from subsystems, and coordinates the inference engines and knowledge bases.

The meta-knowledge base of Meta-COOP consists of such knowledge representations as frame-based, rules-based and method-based and uses the object-oriented programming techniques. Frames are mainly used to describe the static structure of the meta-knowledge base and data structure of objects. The application of frame representation makes the knowledge base have a hierarchical management structure that contains decomposition structure and classification one.

The decomposition structure expresses an 'And' relationship between the current task and all its sub-tasks, that is, the current task can be completed only if all its sub-tasks have been successfully carried out. For example, a fault diagnosis task may be resolved by data acquisition, signal processing and fault analysis. There is a decomposition structure between fault diagnosis and its three sub-tasks.

The classification structure is similar to an 'Or' relationship of the structure. As an example, we can borrow an umbrella 'Or' borrow cash and buy an umbrella. As an engineering example, we can choose a diesel engine or an electric motor in designing a new machine. Users can employ the decomposition or classification relationships to establish their own meta-knowledge base.

The inheritance relationship in Meta-COOP makes the knowledge bases gradually acquire further details, thus, knowledge models can be invoked by their subclasses. As a result, we can enhance the capacity of knowledge representation and efficiency of knowledge processing, and improve the establishment and maintenance of knowledge bases.

Rules represent the heuristic knowledge from domain experts. Classification of rules depends on the domain-specific knowledge and each subset of rules, as a slot value of a frame, is attached to a specified rule slot. The subset can only be invoked by a current frame or its subclasses. Generally speaking, the action (then) part of a rule evaluates attribute values or assigns the values to a specified slot. In addition, a rule can also determine the reasoning path, i.e. send a message to other methods. It should be noted that the control strategy of the rule-based inference engine used is available only in a frame, while the general control strategy of Meta-COOP is determined by the process of message-sending.

The process of message-sending can be viewed as the highest-level manipulation among all inference activities. Its objective is to choose an effective reasoning strategy and suitable knowledge to solve a specified problem, rather than to evaluate a special value.

A method is a kind of procedural language, including a set of operations and statements. A method in Meta-COOP is written in the simplified standard Pascal language syntax. Method can perform numerical computation, functional operations, as well as the process of message-sending. As a message pattern matching process, a method is triggered when the pattern of information sending matches the keywords in this method successfully. The method body is manipulated by the method-based inference. It also supports the message-sending statements, and activates other method bodies in a recursive manner.

In solving a problem, several inference engines in Meta-COOP are invoked from each other. For instance, when the core inference engine performs message transmission, the frame-based inference engine is invoked to manipulate and to evaluate all method slots of an object. Then the core inference engine performs pattern-matching between the message-sending and the keywords in a method body. Finally, the method-based inference engine evaluates the statements in the triggered method body. In addition, the frame-based inference engine may be invoked when the method-based inference engine performs an evaluation operation on a frame (or an object).

The core inference engine also computes the address pointer of a rule subset using the frame-inference engine and executes rules using the rule-based inference engine when there are the statements to drive the rule-based inference engine in a method body. As previously stated, a rule may include the statements to manipulate a frame. The rule-based inference engine deals with the statements by invoking the frame-based inference. Generally speaking, problem-solving in Meta-COOP provides the alternative references for the inference engines and their conflicting coordination through its core inference engine. The reference relationship among these inference engines is shown in Figure 3.6.

Basically, the core inference engine performs message transmission and coordination among other inference engines, which is a distinct difference from other symbolic reasoning systems. The core inference engine integrates various problem-solving methods (inferences) and knowledge representations, thus making Meta-COOP a very powerful method to control integrated distributed intelligent systems.

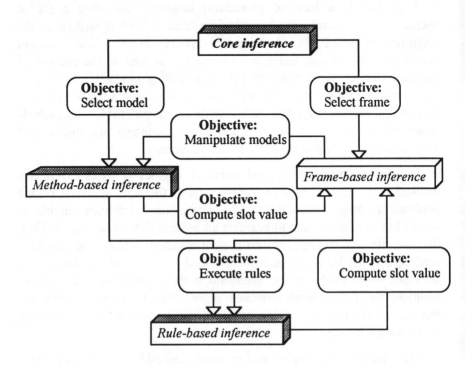

Figure 3.6 Reference relationships among inference engines.

The algorithm of message transmission is a pattern recognition and matching problem. It finds the search area of pattern cognition with the frame-based inference engine, then triggers a method body through method-based inference. Therefore, the algorithm plays an important role in coordinating other inference engines.

The main function of the core inference engine is to coordinate other inference engines. When the core inference engine performs message-sending, a message and the object that receives the messages have to be specified. Of course, the object must include a method body or inherit a method from its ancestral frame. The final result of reasoning does not depend on the method name and the slot name of method, but on the message's content to be sent. An algorithm of message transmission is given below. Its program control flow diagram is demonstrated by Figure 3.7.

Algorithm of message transmission:

(1) Give a message and an object (or frame).

(2) Perform the syntactic analysis to the message.

(3) Search all method slots of the object (or frame).

(4) Carry out evaluation operation on the slots with the frame-based inference engine.

(5) Put the slot values (e.g. the addresses of method) into a queue.

(6) Take a method address from the queue.

(7) Find out the real address of the method.

(8) Find out the address of the information table in the method.

(9) Set $i = 1$.

(10) Take the ith element from the information table as X_i.

(11) Take the ith word from message-sending table as Y_i.

(12) Compute T_i (the type of X_i).

(13) Check whether T_i is a keyword. If yes, then continue. Otherwise, assign $X_i = Y_i$, then go to (6).

(14) Check whether X_i is equal to Y_i. If yes, continue. Otherwise, go to (6).

(15) Compute whether $X_i + 1$ is empty. If empty, continue. Otherwise, set $i = i + 1$, then go to (10).

(16) Compute whether $Y_i + 1$ is empty. If yes, continue. Otherwise, go to (6).

(17) Compute the address of method body.

(18) Send the address to method-based inference engine and execute it.

(19) End.

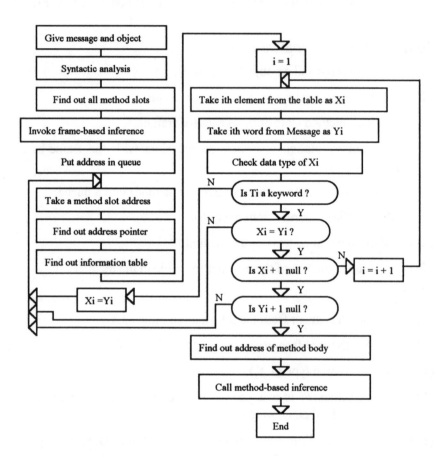

Figure 3.7 Program control flow diagram of message transmission.

On the other hand, the core inference engine also provides a special function to alternately use various inference engines. The algorithm for alternately calling a set of rules is more complicated than the others. During operation, the core inference engine first invokes the frame-based inference engine to find out the address of a set of rules stored in a frame. Then, the rule-based inference engine is invoked to execute reasoning. The following is a description of the algorithm (Figure 3.8).

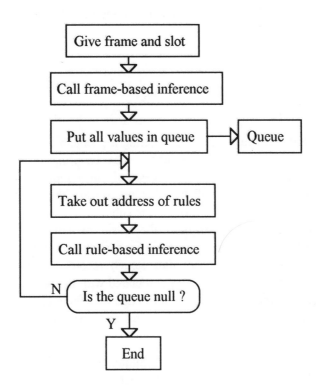

Figure 3.8 Program control flow diagram for calling a set of rules.

Algorithm for alternately calling a set of rules:

(1) Give a frame and a slot.

(2) Determine all values of the slot in the frame (e.g. the address of a set of rules) with the frame-based inference engine. There may be more

than one value in the slot in terms of the inheritance.

(3) Put all values in the queue.

(4) Take out an address of a set of rules from the queue.

(5) Send the address to the rule-based inference engine and invoke it.

(6) Check whether the queue is null. If not, go to (4). Otherwise, continue.

(7) End.

3.12 Interpretation of problem-solving processes

In Meta-COOP, the purpose of interpreting a problem-solving process is to enhance the software transparency, thus enabling users to better understand the decision-making process of an intelligent system. In addition, the interpretation will help beginners learn the specific domain knowledge. The interpretation module (or subsystem) is closely related to the inference engines. Here, only the basic explanations about the reasoning path and computing formula in a method are discussed. The interpretation uses the nodes (or data items) recorded in the reasoning process and the relative information associated with the nodes, backtracks reasoning paths, and displays the knowledge and facts (evidences) used. Of course, some evidences and conclusions, as technical terminology, can be explained, too.

The node structure is an important characteristic of Meta-COOP and provides a list to record the relative arguments that involve the node. The so-called relative argument represents the needed knowledge that triggers the node (e.g. rules and methods) as well as the premises for using the knowledge. Nodes are different from common data, and can record more information. A node not only expresses the conclusion generated, but also includes the knowledge and the facts. When a user requests explanation of the conclusion or data item (node), the interpretation subsystem can automatically display the used knowledge and the premises that activate the node according to the relative arguments, because it records all information to interpret problem solving. A rule syntax is given as an example:

Rule 1:

    Fact   T_requirement range < 70   (A)
    Then Engine_type = "Solid_rocket_engine"   (B)

If fact A (T_requirement range < 70) is provided by user, conclusion B (Engine_type = "Solid_rocket_engine") is obtained. Here, the node structure of conclusion B may be expressed as:

Engine_type = "Solid_rocket_engine"
Rule 1 (*rule number*)
T_requirement range = 40

The node structure of fact A can be shown as:

T_requirement range = 40
USER

The above node structure represents the fact provided by the user. Its relative arguments comes from the user.

The task of setting up a node structure is automatically performed by the inference engines. In Meta-COOP, a new conclusion and the associated knowledge are stored in the node structure. The rule-based inference engine loads a rule serial number and the premises of a rule in a generated node structure, while the method-based inference engine only records a method name. When the facts are provided by user, either the rule-based inference or method-based inference will put 'USER' in a node structure as a relative argument.

In order to provide a user-friendly explanation interface, Meta-COOP numbers all facts and conclusions according to the sequence of display. When a user requests the interpretation of a conclusion or fact, he/she just directly provides the conclusion or types the serial number of the conclusion. Of course, the latter is more convenient than the former. The serial number is automatically generated by the core inference engine in

Meta-COOP (Figure 3.9).

Interpretation algorithm:

(1) Enter a data item that may be a serial number (integer) displayed on the explanation window or a variable name (string).

(2) Check whether the data item is an integer. If yes, continue. Otherwise, go to (4).

(3) Find out the variable name that corresponds to the serial number.

(4) Take the variable value from a node structure in terms of the variable name.

(5) Display the value to be explained.

(6) Take 'knowledge' (rule or method) related with the value from the node structure.

(7) Check whether 'knowledge' is a rule. If not, then continue. Otherwise, go to (9).

(8) Check whether 'knowledge' is a method. If not, continue. Otherwise, print out a method body.

(9) Take the premises of rule from the node structure.

(10) Number all the premises.

(11) Put the numbered premises in the fact list.

(12) Display the rule on an explanation window.

(13) Display the knowledge source.

(14) If the user requests to continuously interpret, then go to (1). Otherwise, continue.

(15) End.

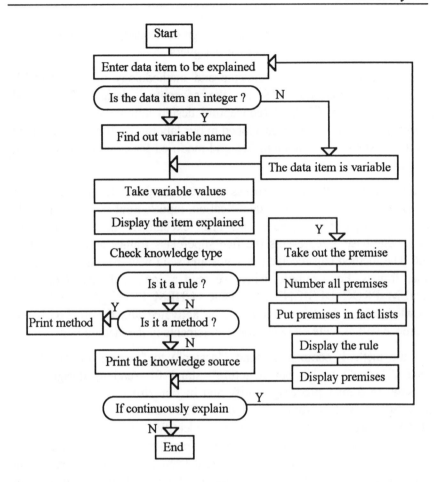

Figure 3.9 Program control flow diagram of interpretation.

3.13 Case study

Meta-COOP has been used in developing integrated distributed intelligent systems. Several applied IDISs have been successfully employed to solve complicated problems in engineering design and manufacturing. As an example, an IDIS for the conceptual design of missile weapon systems, with hundreds of rules and frames, has been developed, which can provide several candidate design schemes for designers.

To better understand the concepts and methodology of Meta-COOP, here we introduce an IDIS for the conceptual design of chemical processes, namely CDACP (conceptual design automation of chemical processes), which has been developed in our laboratory. It should be noted that the system has been simplified and modified for explicitly expressing the use of Meta-COOP.

In CDACP, a chemical process can be divided into five sub-tasks (Douglas, 1988); i.e. process identification, input/output structure, recycle structure, separation system and heat exchanger. In the following, knowledge-base 1 represents the five superclass frames and one method slot. The method can receive messages and send new messages to trigger other frames. For example, when we send a command 'send ("design") "CHEM_PROC_DESIGN"' to Unit CHEM_PROC_DESIGN (i.e. knowledge-base 1), the method will be activated and the method-based inference will execute all operational statements in the method body.

In the method, commands 'send ("design") to "process_identify"' and 'send ("design") to "in_out_structure"' stand for sending 'design' as a message to frames process_identify and in_out_structure, that is to say, activating knowledge-bases 2 and 3. The following two statements '_FRAME.process_iden: = "process_identify"' and '_FRAME.IO_ structure: = "in_out_structure"' drive the frame-based inference engine to put the final results in the slots of current frame, i.e. to put the results in values corresponding to the frame names. So far, all values of slots in knowledge-base 1 have been defined as unknown. In referring to knowledge-base 1, the method-based inference and frame-based inference are alternatively activated.

Knowledge-base 1

```
GLOBE
  Chemprocess: CHEM_PROC_DESIGN;
END
```

Unit: CHEM_PROC_DESIGN in_knowledge_base CDACP.KBS

Memberslot: process_iden from CHEM_PROC_DESIGN
    Inheritance: Override.Values
    Valueclass: Cntns_vs_Batch
    Coordinate.Min: 1
    Coordinate.Max: 1
    Values: Unknown

Memberslot: IO_Structure from CHEM_PROC_DESIGN
    Inheritance: Override.Values
    Valueclass: I/O_Structure
    Coordinate.Min: 1
    Coordinate.Max: 1
    Values: Unknown

Memberslot: Recycle_Structure from CHEM_PROC_DESIGN
    Inheritance: Override.Values
    Valueclass: Recycle_Structure
    Coordinate.Min: 1
    Coordinate.Max: 1
    Values: Unknown

Memberslot: Separation_Sys from CHEM_PROC_DESIGN
    Inheritance: Override.Values
    Valueclass: Separation_Sys
    Coordinate.Min: 1
    Coordinate.Max: 1
    Values: Unknown

Memberslot: Heat_Exchanger from CHEM_PROC_DESIGN
    Inheritance: Override.Values
    Valueclass: Heat_Exchanger
    Coordinate.Min: 1
    Coordinate.Max: 1
    Values: Unknown

Memberslot: design_method from CHEM_PROC_DESIGN
    Inheritance: METHOD
    Valueclass: METHODS
    Coordinate.Min: 1
    Coordinate.Max: 1
    Values: design_method

METHOD design_method (design: keyword)

```
VAR
  Range: string ;
BEGIN
  send ("design") to "process_identify";
  send ("design") to "in_out_structure";
  _FRAME.process_iden:="process_identify";
  _FRAME.IO_structure:="in_out_structure";
END.
```

Knowledge-base 2 checks a chemical process to be designed is either continuous or batch. The frame is a subclass of CHEM_PROC_DESIGN frame and has three slots. Slot process type records the result of identification (continuous or batch). Slot 'process_iden_rules' stores a set of rules (rule 31 and 38) used in identifying a chemical process. Method slot 'process_iden_method' receives a message from other frames and calls the rules-based inference engine for executing the statement 'reason (_FRAME, "process_iden_rules")'.

When knowledge-base 2 is activated by a statement 'frame-name, "method name"', the rule-based inference automatically runs and puts the result in the current frame through processing the statement '_FRAME.process_type = "problem solution"'. Knowledge-base 2 is given as follows:

Knowledge-base 2

```
GLOBE
  process_identify: process_iden;
END
```

Unit: process_iden in_knowledge_base CDACP.KBS

Memberslot: process_type from process_iden
  Inheritance: Override.Values
  Valueclass: String
  Coordinate.Min:1
  Coordinate.Max: 1
  Values: Unknown

Memberslot: process_iden_rules from process_iden
  Inheritance: Override.Values

Valueclass: RULES
Coordinate.Min: 1
Coordinate.Max: 1
Values: {
      rule 31

            fact specification.market_forces="seasonal_production"
            or   specification.market_forces="short_prod_lifetime"
            or   specification.production_rate<=10000000
            or   specification.products="multiple"
            or   specification.reaction_time="short_time"
            or   specification.material_act="slurry"
            or   specification.material_flow="slow"
            or   specification.material_taste="rapidly_fouling"
            then _FAME.process_type="batch_process"
      rule 38

            fact specification.market_forces="annual_production"
            or   specification.market_froces="long_prod_lifetime"
            or   specification.production_rate>10000000
            or   specification.products="single"
            or   specification.reaction_time="long_time"
            or   specification.material_taste="tasteless"
            or   specification.material_flow="fast"
            or   specification.material_act="o_slurry"
            or   specification.material_flow="fast"
            then _FRAME.process_type="continue_process"
            }

Memberslot: process_iden_method from process_iden
  Inheritance: METHOD
  Valueclass: METHODS
  Values: process_iden_method

  METHOD process_iden_method (design: keyword)
VAR
  design:keyword;
BEGIN
  reason (_FRAME,"process_iden_rules");
END.

Similarly, knowledge-base 3 is also a subclass of 'CHEM_PROC_ DESIGN'. The second statement 'send ("design") to "in_out_structure"' in 'design_method' of knowledge-base 1 can trigger knowledge-base 3. Frame 'IO_structure' includes five classification slots that express the classification relationship of input/output structure of a chemical process (static knowledge), which are purify_feed, pro_by_product, gas_recycle, process_reactant and stream_number. Other rule slots store the dynamic knowledge (rules) that can be called by the rule-based inference. In addition, knowledge-base 3 has a method slot to receive and send messages, thus to execute the rule-based inference engine, and to perform numerical models.

Knowledge-base 3

GLOBE
    in_out_structure: IO_structure;
    first_guess: string;
    easier_separation: string;
    impurity_amount: string;
    reversible_byproduct: string;
    RcvCost_mthn_Rawmlst: string;
    the_reactant_is: string;
    the_byproduct_is: string;
    the_impurity_is : string;
END

Unit: IO_structure in_knowledge_base CDACP.KBS

Memberslot: purify_feed from IO_structure
    Inheritance: Override.Values
    Valueclass: String
    Coordinate.Min: 1
    Coordinate.Max: 1
    Values: Unknown

Memberslot: proc_by_product from IO_structure
    Inheritance: Override.Values
    Valueclass: String
    Coordinate.Min: 1
    Coordinate.Max: 1
    Values: Unknown

Memberslot: gas_recycle from IO_structure
    Inheritance: Override.Values
    Valueclass: String
    Coordinate.Min: 1
    Coordinate.Max: 1

Values: Unknown

Memberslot: proc_reactant from IO_structure
  Inheritance: Override.Values
  Valueclass: String
  Coordinate.Min: 1
  Coordinate.Max: 1
  Values: Unknown

Memberslot: stream_number from IO_structure
  Inheritance: Override.Values
  Valueclass: String
  Coordinate.Min: 1
  Coordinate.Max: 1
  Values: Unknown

Memberslot: purify_feed_rules from IO_structure
  Inheritance: Override.Values
  Valueclass: RULES
  Coordinate.Min: 1
  Coordinate.Max: 1
  Values: {
      rule 41
            fact specification.an_impurity_is_in="gas_feed_stream"
            then first_guess="process_impurity"
      rule 42
            fact specification.an_impurity_is_in
            = "azeotrope_with_reactant"
            then first_guess="process_impurity"
      rule 43
            fact specification.feed_impurity_act="inert"
            and easier_separation="yes"
            then first_guess="process_impurity"
      rule 44
            fact specification.feed_impurity_act="no_inert"
            and first_guess="process_impurity"
            then process_method="remove_it";
                _FRAME.purify_feed="remove_it"
      rule 45
            fact impurity_amount="large"
            and first_guess="process_purity"
            then process_method="remove_it";
                _FRAME.purify_feed="remove_it"
      rule 46
            fact specification.feed_impurity_is="catalyst_poison"

```
                            and first_guess="process_impurity"
                            then process_method="remove_it";
                                _FRAME.purify_feed="remove_it"
            rule 47
                            fact specification.an_impurity_is_in="liguid_feed_sream"
                            and specification.stream_has="product"
                            and first_guess="process_impurity"
                            then process_method="use_separation";
                                _FRAME.purify_feed="use_separation"
            rule 48
                            fact specification.an_impurity_is_in="liguid_feed_sream"
                            and specification.stream_has="by_product"
                            and first_guess="process_impurity"
                            then process_method="use_separation";
                                _FRAME.purify_feed="use_separation"
                            }
```

Memberslot: proc_by_prdct_rules from IO_structure
  Inheritance: Override.Values
  Valueclass: RULES
  Coordinate.Min: 1
  Coordinate.Max: 1
  Values: {
```
            rule 52
                            fact reversible_byproduct="yes"
                            and RcvCost_mthn_Rawmlst="yes"
                            then _FRAME.proc_by_product="unnecessary"
            rule 54
                            fact reversible_byproduct="yes"
                            and RcvCost_mthn_Rawmlst="no"
                            then _FRAME.proc_by_product="necessary"
            rule 56
                            fact reversible_byproduct="no"
                            then _FRAME.proc_by_product="necessary"
                            }
```

Memberslot: gas_recyl_pur_rules from IO_structure
  Inheritance: Override.Values
  Valueclass: RULES
  Coordinate.Min: 1
  Coordinate.Max: 1
  Values: {
```
            rule 65
                            fact specification.boilp_of_reactant<-48.
                            then the_reactant_is="light_cmpnt"
```

```
rule 66
            fact specification.boilp_of_impurity<-48
            then the_impurity_is="light_cmpnt"
rule 67
            fact specification.boilp_of_byprofuct<-48.
            then the_byproduct_is="light_cmpnt"
rule 68
            fact the_reactant_is="light_cmpnt"
            and the_impurity_is="light_cmpnt"
            then _FRAME.gas_recycle="gas_recycle_purge"
rule 69
            fact the_reactant_is="light_cmpnt"
            and the_byproduct_is="light_cmpnt"
            then _FRAME.gas_recycle="gas_recycle_purge"
            }

Memberslot: IO_structure_method from IO_structure
  Inheritance: METHOD
  Valueclass: METHODS
  Values: IO_structure_method
  Coordinate.Min: 1
  Coordinate.Max: 1
  Values: IO_structure_method

METHOD IO_structure_method (design: keyword)
VAR
  design: keyword;
BEGIN
  reason (_FRAME,"purify_feed_rules");
  reason (_FRAME,"proc_by_prdct_rules");
  reason (_FRAME,"gas_recyl_pur_rules");
END.
```

Knowledge-base 4 can be viewed as a database that records the static data and is connected with the window of the user interface. Interactively, data and information from users are automatically stored in knowledge-base 4, and referred to by the frame-based inference, rule-based inference and method-based inference.

Knowledge base-4:

GLOBE
  specification: static_facts;
END

Unit: static_facts in_knowledge_base CDACP.KBS

memberslot: market_forces from static_facts
  Inheritance: Override.Values
  Valueclass: String
  Coordinate.Max: 1
  Coordinate.Min: 1
  Values: annual_production

memberslot: production_rate from static_facts
  Inheritance: Override.Values
  Valueclass: Real
  Coordinate.Min: 1
  Coordinate.Max: 1
  Values: 20,000,000.

memberslot: products from static_facts
  Inheritance: Override.Values
  Valueclass: string
  Coordinate.Min: 1
  Coordinate.Max: 1
  Values: single

memberslot: reaction_time from static_facts
  Inheritance: Override.Values
  Valueclass: string
  Coordinate.Min: 1
  Coordinate.Max: 1
  Values: long_time

memberslot: material_taste from static_facts
  Inheritance: Override.Values
  Valueclass: string
  Cardinality.Min: 1
  Coordinate.Max: 1
  Values: tastenless

memberslot: material_flow from static_facts
  Inheritance: Override.Values
  Valueclass: string
  Coordinate.Min: 1
  Coordinate.Max: 1
  Values: fast

memberslot: material_act from static_facts
  Inheritance: Override.Values
  Vaueclass: string
  Coordinate.Min: 1
  Coordinate.Max: 1
  Values: inert

memberslot: feed_impurity_act from static_facts
  Inheritance: Override.Values
  Valueclass: string
  Coordinate.Min: 1
  Coordinate.Max: 1
  Values: inert

memberslot: an_impurity_is_in from static_facts
  Inheritance: Override.Values
  Valueclass: string
  Coordinate.Min: 1
  Coordinate.Max: 1
  Values: gas_feed_stream

memberslot: feed_impurity_act from static_facts
  Inheritance: Override.Values
  Valueclass: string
  Coordinate.Min: 1
  Coordinate.Max: 1
  Values: inert

memberslot: feed_impurity_is from static_facts
  Inheritance: Override.Values
  Valueclass: string
  Coordinate.Min: 1
  Coordinate.Max: 1
  Values: catalyst_poison

memberslot: stream_has from static_facts
  Inheritance: Override.Values
  Valueclass: string

Coordinate.Min: 1
Coordinate.Max: 1
Values: product

memberslot: boilp_of_byprofuct from static_facts
   Inheritance: Override.Values
   Valueclass: real
   Coordinate.Min: 1
   Coordinate.Max: 1
   Values: -50

memberslot: boilp_of_reactant from static_facts
   Inheritance: Override.Values
   Valueclass: real
   Coordinate.Min: 1
   Coordinate.Max: 1
   Values: -60

memberslot: boilp_of_impurity from static_facts
   Inheritance: Override.Values
   Valueclass: real
   Coordinate.Min: 1
   Coordinate.Max: 1
   Values: -70

When a user types CDACP and a password under the UNIX™ operating system, the user-interface appears on the screen as shown in Figure 3.10. The interface includes five perpetual windows and several temporary windows to provide interpretation and help.

The window at the left top corner gives users several icons and information (Figure 3.11). The icons are activated by a mouse and execute various functions such as quitting from the system, interpretation, reasoning and introduction of the system. The data item 'Message:' expresses a message to be sent and 'Object:' stands for an object to receive the message. The window on the right side is the menu that includes several function buttons (Figure 2.9). The user can trigger each button by a mouse to execute the functions. For example, if the button [FILES] is typed, the temporary window choosing data files and knowledge bases will appear on the screen as shown in Figure 3.12.

The window on the left-hand side is a display area that demonstrates the overall process of problem-solving as shown in Figure 3.13. If the user needs Meta-COOP to explain the data item or reasoning result, he/she can press the icon 'WHY' and specify the data item to be explained. Then, the content of interpretation (such as rules and methods) will be shown on the window in the middle of the screen. For instance, when data items 2 and 1 are specified by the user, rule 46 and rule 52 that have been used will be displayed in the window (Figure 3.14). Finally, the widow at the bottom is a command window to execute UNIX™ operating commands and to perform information exchange between user and UNIX™ operating system under the Meta-COOP environment.

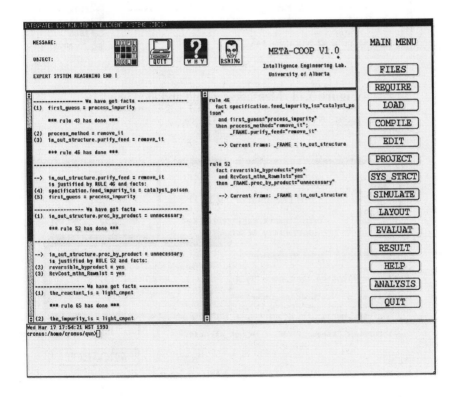

Figure 3.10 User interface of Meta-COOP.

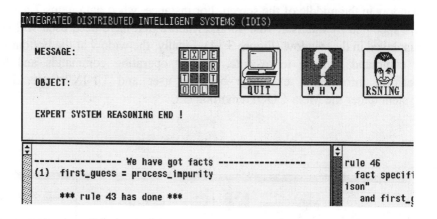

Figure 3.11 Message-sending and icon window.

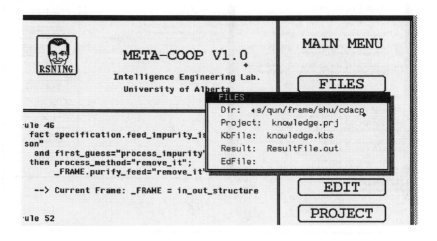

Figure 3.12 Temporary window for file selection.

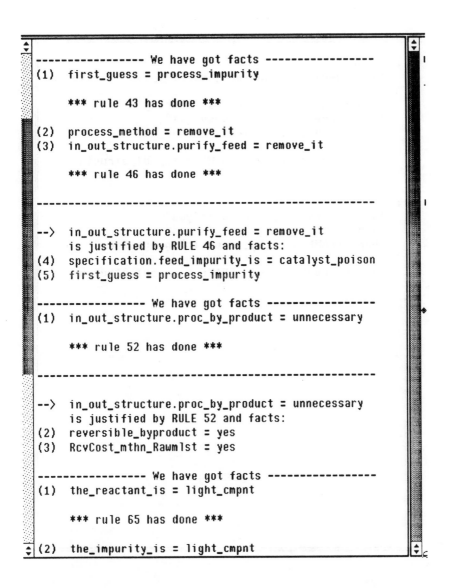

```
----------------- We have got facts -----------------
(1)   first_guess = process_impurity

      *** rule 43 has done ***

(2)   process_method = remove_it
(3)   in_out_structure.purify_feed = remove_it

      *** rule 46 has done ***

-----------------------------------------------------

-->   in_out_structure.purify_feed = remove_it
      is justified by RULE 46 and facts:
(4)   specification.feed_impurity_is = catalyst_poison
(5)   first_guess = process_impurity
----------------- We have got facts -----------------
(1)   in_out_structure.proc_by_product = unnecessary

      *** rule 52 has done ***

-----------------------------------------------------

-->   in_out_structure.proc_by_product = unnecessary
      is justified by RULE 52 and facts:
(2)   reversible_byproduct = yes
(3)   RcvCost_mthn_Rawmlst = yes
----------------- We have got facts -----------------
(1)   the_reactant_is = light_cmpnt

      *** rule 65 has done ***

(2)   the_impurity_is = light_cmpnt
```

Figure 3.13 Display window of a problem-solving process.

```
rule 46
  fact specification.feed_impurity_is="catalyst_po
ison"
   and first_guess="process_impurity"
   then process_method="remove_it";
        _FRAME.purify_feed="remove_it"

  --> Current Frame: _FRAME = in_out_structure

rule 52
  fact reversible_byproduct="yes"
   and RcvCost_mthn_Rawmlst="yes"
   then _FRAME.proc_by_product="unnecessary"

  --> Current Frame: _FRAME = in_out_structure
```

Figure 3.14 Interpretation window of Meta-COOP.

4

Integrated distributed intelligent design environment

4.1 Conceptual design of mechanical products

In this chapter, we first introduce the fundamentals and characteristics of concept design. Then, a general problem-solving strategy and methodologies to implement conceptual design automation are proposed. An integrated distributed intelligent design environment (IDIDE) for building conceptual design expert systems is presented. Its fundamental principles, system organization and implementation techniques are discussed. Finally, two industrial applications are presented.

4.1.1 Conceptual design

Conceptual design is a creative activity and an important decision-making process during overall product design intended to reduce energy consumption, obtain more profits, reduce environmental effects of effluents, and ensure flexibility, operability, manufacturability and safety of manufacturing processes. Therefore, the quality of conceptual design plays a key role in determining the final quality of products and the profit of plants.

Development of industrial technology is closely related to applications of computers. In engineering design, computer-aided design (CAD) technology has evolved into a new generation of design techniques. It has also paved the way for the implementation of computer-integrated manufacturing (CIM) systems. However, the key issue in accomplishing CAD is the conceptual design automation of industrial products (Wang *et al.*, 1990a).

Conceptual design is an important but difficult target in CAD. Product quality, reliability, production cost and productivity depend not only on the quality of the detailed design and manufacturing process for each component or part, but also heavily on the conceptual design and the coordination of layout design of components (i.e. the quality of design synthesis). Actually, the quality of conceptual design can determine the nature of a product and its competitiveness in the marketplace.

In the past few decades, computers have been extensively used in optimization, finite element analysis, reliability design, computer graphics and simulation for detailed design in engineering, but there has been relatively little achievement at the conceptual design stage. The use of computers has been limited almost exclusively to purely algorithmic solutions. Because no automation is available for conceptual design (design synthesis), the development and applications of CAD technology have been hindered.

The conceptual design of mechanical products consists of scheme design and layout design. Scheme design conceptually determines specifications, performances, functions and structure parameters of a product, and selects its structural forms, materials and configuration, etc. All conceptual design results provide numerical and symbolic information for the follow-on detailed design. Layout (or structure) design places all parts and components. Clearly, these two aspects usually involve ill-structured problems, which consist of non-numerical or non-algorithmic information, and are not amenable to purely algorithmic computation (Rao, *et al.*, 1989). The methodology to solve these ill-structured problems is understanding, defining, reasoning and decision-making based on specific domain knowledge and expert experience. According to these characteristics, conventional CAD techniques cannot meet the needs of conceptual design. Intelligent system technology is suggested as an alternative to solve the problems.

4.1.2 Characteristics of conceptual design

Conceptual design is a complex engineering task involving performing numerical computation and symbolic inference as well as graphics simulation. As reported recently, many practical expert systems have been developed for manufacturing engineering. Most of them are used in fault diagnosis and production planning, and few are used in engineering design. In addition, hundreds of the tools (or shells) for building expert systems are available in the software marketplace, many of them available only for the special applications of diagnosis and planning, and not suitable for design. It is a very difficult task to develop intelligent systems for engineering design due to the following problems:

(a) Multiplicity of design results and uncertainty of objectives

Usually, problem-solving in diagnosis involves a 'multiple input/single output' problem-solving pattern; that is, one conclusion (output) can be inferred from some evidences with an inference engine (sometimes more than one). In contrast, design problem-solving is 'single input/multiple output'; that is, a few results which can meet the same requirements may be obtained at the same time. As a result, two obstacles may be encountered in engineering design: large decision space and comprehensive evaluation of designs. The problems we face here are how to find all acceptable schemes that satisfy the design requirements and how to select the best one from these acceptable schemes.

(b) Multiple levels and multiple objectives of design tasks

Obviously, a design needs to perform various sub-tasks that lie on different levels. For instance, the design of machine tools involves many aspects such as transmission, hydrostatic circuit, electric circuit, power utility, operation and so on. They may be implemented on different levels and controlled by meta-knowledge (Rao *et al.*, 1989). Thus, new problems arise: how to automatically resolve and plan a design task, how to represent the relationship among sub-tasks, how to solve conflicts, and how to choose an appropriate problem-solving strategy to match a sub-task.

(c) Intelligent design environment for computation and inference

Conventional expert systems emphasize symbolic processing and non-algorithmic inferences. Because problem-solving in design needs not only symbolic reasoning but also numerical calculation, the integrated distributed intelligent design environment should be able to use a variety of existing analysis and simulation packages, and to exchange information with a database management system at any time.

(d) Multiplicity of knowledge representation and problem-solving strategy

Product design deals with various problem-solving methods and knowledge representation forms. For example, it often employs reasoning, calculation, table look-up and graphics. During the development of such intelligent systems, the knowledge base, database, and control strategy must be segregated to allow users to organize the different models and domain expertise efficiently because each of these components can be designed and modified separately.

(e) Structure problem-solving and geometry knowledge representation

The final results of product design, including those from conceptual design and detailed design stages, should be ultimately represented by drawings that involve 80% of design information. The design deals with complicated geometric information, and implements structural and layout designs that touch upon the representation of geometric information and inference of space knowledge. The challenge here is how to describe and cope with geometric, functional and topological information of products.

(f) Complexity of redesign

Redesign is the inevitable obstacle to solving design problems. When the results are unsatisfactory to users, an integrated distributed intelligent design environment (IDIDE) has to carry out redesign. Obviously, with increase of system size and problem complexity, redesign will be very difficult. There exist such difficulties as how to store and apply the failure information to directing redesign, how to decide an optimum problem-solving strategy when multiple tasks conflict, and how to implement expertise knowledge.

4.2 DAER model

4.2.1 Design–analysis–evaluation–redesign model (DAER)

'Design–analysis–evaluation–redesign' (DAER) is a very effective and practical model for the conceptual design of mechanical products. It summarizes the current engineering design methodologies (Dixon and Simmons, 1983). The DAER model reflects the expert's reasoning activity in solving engineering design problems and can well handle design problems that need to employ empirical knowledge. IDIDE is developed using a DAER model.

Engineering design is actually an art and a creative process (Douglas, 1988). Designers might try to approach design problems in much the same way as a painter develops a painting. In other words, designers' original design procedures should correspond to the development of a pencil sketch, where designers want to suppress all but the most significant details of the design; i.e. they want to discover the most important parts of a product to be designed, which determine the final performance and price of a product. An artist next evaluates the preliminary painting and makes modifications, using only gross outlines of the subjects. Similarly, designers want to evaluate their first guess at a design and generate a number of design alternatives. In this way, designers hope to generate a 'reasonable-looking' rough product design before they start adding more details.

To sum up, a design procedure can be divided into two stages: (1) generating a preliminary design scheme, and (2) repeating the procedure of 'analysis evaluation redesign' until all requirements and constraints are met (i.e. all the design results are feasible). As shown in Figure 4.1, the DAER model puts forward a preliminary design scheme (synthesis) based on customer's requirements and a marketing investigation. Then, all schemes are analyzed and evaluated. Finally, the scheme that best meets the acceptability requirements is selected from all candidates, based on design criteria. If no design scheme can be accepted, then redesign will be performed.

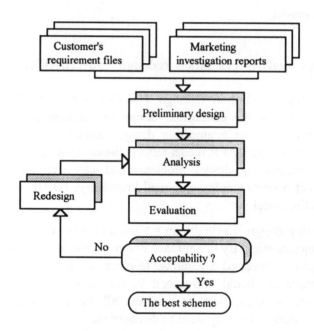

Figure 4.1 DAER model.

(a) Preliminary design

Preliminary design usually appears in a new product design. Its problem-solving strategy might include the following processes:

- generating a design scheme on the basis of comparison with existing design results of products;
- generating a design scheme that meets all requirements and constraints;
- generating a design scheme that only meets part of the constraints; and

- generating a design scheme randomly.

These methods can be used independently or in combination. The preliminary design relies on specific domain knowledge and experts' experience. It may be neglected that some designs are in the redesign area to improve old products. In this case, the preliminary design results are equal to the parameters of an existing products. These results are directly analyzed and evaluated without loading the preliminary design module, because the results already include enough data and information for the analysis module.

(b) Analysis

Engineering analysis can provide more detailed data and information so that an evaluation procedure can be executed. In general, two types of information are produced in the design process: one is symbolic information, and the other is numerical information. Handling numerical information has to use such computer-aided engineering techniques as finite element analysis, optimization and reliability analysis.

(c) Evaluation

The evaluation of design schemes involves a comprehensive examination of all design targets to choose an optimal design scheme. It requires not only statistical models but also experts' experience.

(d) Acceptability

This is a 'Yes/No' test to determine whether all design requirements and constraints have been met.

(e) Redesign

Redesign needs to process various feedback messages from the earlier analyses and evaluations, then, generate a modified design scheme based on the feedback information.

4.2.2 Definitions

To better understand and use the DAER model, a general problem-solving strategy based on the DAER model is discussed. First of all, it is necessary to introduce a few definitions and terminology.

Definition 1. A concept is defined as an abstract description of the natural property of an object. Each concept has its own identifier, referred as C_i.

Definition 2. Concept space is a set that includes all concepts in a specific domain. These concepts are organized in a specific order and hierarchy to define the relationships among them. Concept space is written as *CS*, where, $C_i \in CS$.

Definition 3. Function concept is an abstract description of the features of system functions. It is expressed as FC_i.

Definition 4. Function concept space is a set that includes all functions in a specific domain. It is written as *FCS*. Similarly, $FC_i \in FCS$, and $FCS \in CS$.

Definition 5. Structure concept is an abstract description of the system structures. It is represented as SC_i. There exists such a relationship that $SC_i \in CS$.

Definition 6. Structure concept space consists of all structure concepts. It is also a subset of *CS*. It is denoted as *SCS*. $SC_i \in SCS$, and $SCS \in CS$.

Definition 7. Effective concept is the concept that satisfies the application environment and objectives as well as key constraints. Its notation is EC_i.

Definition 8. Effective concept space (*ECS*) consists of all effective concepts. $EC_i \in ECS$, and $ECS \in CS$.

Definition 9. Effective function concept is a function concept that satisfies the application specifications and constraints. It is denoted as EFC_i.

Definition 10. Effective function concept space (*EFCS*) includes all effective concepts. $EFC_i \in EFCS$, while $EFCS \in FCS$.

Definition 11. Effective structure concept (*ESC*) is a structure concept that satisfies the application environment, objectives, constraints and effective function concepts. $ESC_i \in SCS$.

Definition 12. Effective structure concept space (*ESCS*) consists of all effective structure concepts. $ESC_i \in ESCS$, and $ESCS \in SCS$.

Definition 13. Design pattern is a tree structure that consists of nodes and arcs. Each node represents an effective structure concept, and each arc indicates an 'AND' relation of nodes, or an 'OR' relation of a single node. It is denoted as DP_i.

Definition 14. Design pattern set (*DPS*) is equivalent to *ESCS*. Each design pattern represents a design scheme. Therefore, a pattern set is also a scheme set that meets the application environment and design specifications. $DP_i \in DPS$.

4.3 Problem-solving strategy

A general problem-solving strategy of conceptual design based on the DAER model is described by the following five stages (Figure 4.2).

Stage 1 is a problem definition stage for design tasks (transferring the application environments and design purposes to functions). Functions to be used are chosen from the expertise function memory (which can be viewed as a part of the knowledge base) in terms of the application environments and objectives provided by the users. For example, site location has an impact on the conceptual design of chemical processes because the utilities available on site (such as cooling water temperatures) will depend on the geographical location. The knowledge to define functions is shallow knowledge (Kapp, 1989).

The first step in problem solving can be expressed as follows:

$$\text{STEP } 1 = (FCS, \sum_{i=1}^{n} EFC_i \,|\, S_i \text{ and } T_i, EFCS) \qquad (4.1)$$

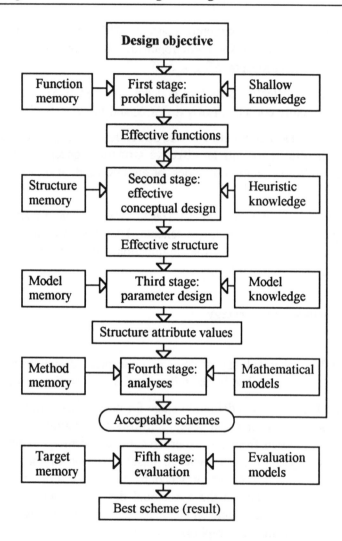

Figure 4.2 Problem-solving strategy.

where, S_i and T_i (i = 1, 2, ... n) represent specifications and constraints provided by users. The objective in the above expression is to find an EFC_i to satisfy S_i and T_i in FCS, then to combine these EFC_i into $EFCS$.

Stage 2 is an effective conceptual design stage (from functions to structures). The structures to execute functions are selected from a structure memory (it can be viewed as another part of the knowledge base). The communication between functions and structures is not a 'one-to-one' mapping. Such a 'multiple-to-multiple' mapping configuration (Figure 4.3) indicates that a function can be realized with many different structures, and a structure may possess many functions. For example, the function for cooling can be implemented by several structures: cooling water, air, oil or others. This 'multiple-to-multiple' mapping makes the pattern design alternatives more complicated and diversified. Each design alternative has a design scheme. If there is more than one design alternative (pattern), an optimal or near optimal solution must be selected. In the case where no design alternatives are available, new design techniques must be used (since no existing structure can be used for the needed functions). If the existing design alternatives fail to satisfy application requirements, the design has to be improved. The knowledge to formulate effective structural concepts is heuristic knowledge that can be represented by heuristic rules.

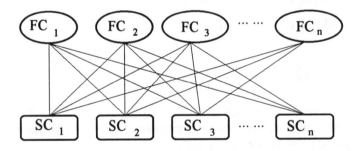

Figure 4.3 Configuration of multiple to multiple mapping.

The following expression is the second step of problem-solving:

$$\text{STEP 2} = (SCS, \sum_{i=1}^{n} ESC_i \,|\, S_i \text{ and } T_i, ESCS) \tag{4.2}$$

This expression presents how to select all ESC_i that satisfy S_i, T_i and EFC_i in SCS, then to combine these ESC_i into $ESCS$.

In general, $ESCS$ is a set of design patterns (alternatives). It is necessary to resolve them into individual patterns such that each individual pattern is one design scheme. The process is called by the scheme-resolving module. It can be expressed as the following algorithm:

$$\text{RESOLVING} = (ESCS, OP, \sum_{i=1}^{n} DP_i) \qquad (4.3)$$

where OP stands for a set of resolving operations. The purpose of the above expression is to divide $ESCS$ (or DPS) into several DPs (or design schemes) with OP.

Stage 3 is a parameter design stage (from structures to parameters). At this stage, the detailed description of structures can be completed by using the design models stored in model memory according to the characteristics of effective structural concepts. The knowledge to determine structural attributes is deep knowledge, namely model knowledge, which is represented by object-oriented frames.

The third step can be summarized below:

$$\text{STEP 3} = (\sum_{i=1}^{n} DP_i, OPF, \sum_{i=1}^{n} ADP_i) \qquad (4.4)$$

where OPF represents an operating set of frames, and ADP_i is a design pattern with attributes and values (parameters). The expression indicates that it converts DP_i into ADP_i by OPF.

Stage 4 is an analysis stage (from parameters to analyses). Since functions and structures share a 'multiple-to-multiple' mapping configuration, numerous design schemes are usually produced. After parameters are given, all design schemes will be analyzed by numerical computation methods (e.g. statistical analysis, optimization, and so forth) from method memory. Conventional CAD techniques can be utilized here.

The fourth stage of problem solving can be expressed as:

$$\text{STEP } 4 = (\sum_{i=1}^{n} ADP_i, \ OM, \ \sum_{j=1}^{m} ADP_j) \qquad (4.5)$$

where *OM* represents an operating set of analysis methods, and $n \geq m$. The algorithm is used to analyze every design scheme so that the feasible schemes are selected from ADP_i to satisfy the requirements of analyses. Usually, a few schemes (or patterns) are omitted, and only the practical schemes are kept.

Stage 5 is a final stage for comprehensive evaluation (from analysis to evaluation). Based on the analysis data, a proper evaluation target system from target knowledge memory and a comprehensive mathematical model from evaluation models is chosen to evaluate the selected practical schemes. Techniques of fuzzy mathematics and system engineering are used in evaluation.

The fifth step can be represented as:

$$\text{STEP } 5 = (\sum_{j=1}^{m} ADP_j, \ OE, \ ADP^*) \qquad (4.6)$$

where *OE* is an operating set of evaluations. This algorithm intends to find the best scheme among all the practical candidate patterns by using evaluation *OE*. *ADP** is an optimal design to be sought.

Each stage in the problem-solving strategy is very important. It combines numerical calculation (such as mathematical modeling, optimization, and scheme analysis) with symbolic reasoning (knowledge representation and model handling as well as scheme evaluation) to accomplish the objectives in every stage.

4.4 System configuration

A good problem-solving strategy must match a good control structure to ensure software quality and efficiency. The control structure of IDIDE (Figure 4.4) has been developed on the basis of the DAER model and meta-system architecture (Rao, 1991). It can be used as a general framework for developing integrated distributed intelligent systems for accomplishing the conceptual design of mechanical products. The control structure uses modular techniques. It consists of a menu, fifteen subsystems (or modules), and several different knowledge bases (Figure 4.4). The following paragraphs briefly describe these subsystems.

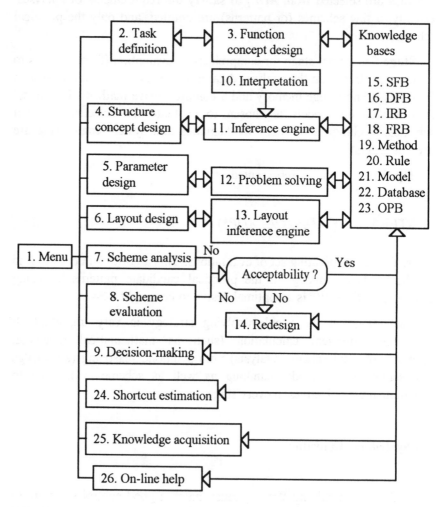

Figure 4.4 Software configuration of IDIDE.

1. **Menu management system:** This can guide users to select modules and observe performances. Because the menu system employs a tree structure (each subsystem has its own sub-menu) and an object-oriented programming technique for man–machine interface, users can select and run modules according to the contents of the screen.

2. **Task definition module:** This is a window interface to input information. Users can define design tasks, the working environment of applications, purposes, and specifications through the window.

3. **Function concept design module:** This functions as the first step of the problem-solving strategy. Its purpose is to further expand facts and information in the knowledge base based on the existing specifications, to determine the global variables that can be shared by many subsystems. In general, global variables are not well understood. The entire system must be considered when determining the global variables. However, this is a difficult task for a new designer. In addition, a user can never know whether he/she receives correct and complete information. If some important data are missed, the module should supply these data using domain-specific knowledge.

4. **Structure conceptual design module:** This can select the types of machines and potential structural configuration to satisfy the specified functions and key constraints. This module works in the second step and provides a parameter design with a variety of features. Its knowledge base includes experts' experience and heuristic knowledge, which are represented as heuristic rules. Inference employs constraint reasoning (Wang et al., 1990a).

5. **Parameter design module:** Its function is similar to the third step of design. With the structures obtained from concept design and the facts provided by users, attribute values (parameters) of a mechanical system or product are determined. Parameter design usually deals with local variables by shortcut calculations. Local variables are usually well understood by engineers, because these variables are only associated with a subsystem, rather than with the overall system or other subsystems. The knowledge of parameter design is expressed by the data structure of an object-oriented frame that is operated by the problem solver.

6. **Layout design module:** Based on the geometrical parameters, the layout design module determines the position and orientation of each component and makes a drawing design. Its other purpose is to perform a simulation for testing geometric interference among bodies.

7. **Scheme analysis module:** This functions in the fourth step. Based on the results of parameter design and layout design, this module analyzes each preliminary design scheme and provides data for evaluation. There are two paths in the module. If there is a more satisfactory scheme, the system selects the next module. Otherwise, local redesign is required. In addition, the module can call the existing analysis and simulation software packages.

8. **Scheme evaluation module:** This module is used in the fifth step of problem-solving. Its purpose is to evaluate comprehensive functions. In other words, all schemes entered in the evaluation module are practical ones that are only different from each other in quality. The evaluation uses the comprehensive evaluation models, fuzzy mathematics and system engineering techniques (Wang *et al.*, 1989a; Zhou *et al.*, 1989b). Indices (or targets such as material price and manufacturing cost) and weights are selected by domain experts. There are two paths in the evaluation module. If the evaluation results satisfy the specifications provided by the users, the information is sent to the decision-making module. Otherwise, a global redesign is performed.

9. **Decision-making module:** In general, multiple schemes are generated as the result of a design. During a mechanical product design, the index system solicits various opinions from different domain experts. Thus, conflicting solutions are generated from the different experts' opinions. The decision-making module picks the best scheme from among these solutions.

10. **Interpretation module:** This is connected with the inference engine and provides the interpretation for concepts and reasoning paths to help users to understand items and concepts, as well as to manage the system.

11. **Inference engine:** The inference engine usually performs specific tasks and executes the system based on the fired data and knowledge

representation. To solve large and complex engineering design problems, a single inference engine that provides only one reasoning technique is unsatisfactory. The integration of different inference mechanisms is very useful in solving real-world problems.

12. **Problem solver:** This operates a method knowledge base with a variety of problem-solving strategies, including reasoning, calculation, table look-up, curve observation and analogy. In conceptual design, these engineering design methods are often used together. For example, reasoning determines empirical coefficients, table-look and curve observation provide numerical information, and analogy helps designers to choose better structure types.

13. **Layout inference engine:** Layout inference is much more complicated than symbolic inference because designing this inference engine deals with the shapes, positions, orientations of a body to be assembled, as well as space constraints and interference of bodies.

14. **Redesign module:** Redesign uses the messages from the earlier analysis and evaluation as its input to improve the design quality and to find new design solutions based on feedback information.

15. **Static facts base (SFB):** This stores task definitions and specifications provided by users. The information in the SFB contains the essential conditions (constraints) for functional and structural concept design. The facts in the SFB are expressed as *vector lists*.

16. **Dynamic facts base (DFB):** In a reasoning process, users must continuously provide more detailed facts and data, which are stored in the DFB. In IDIDE, the DFB is only associated with the inference engine and supports the interpretation module.

17. **Intermediate result base (IRB):** This stores intermediate results during processes of symbolic reasoning and numerical computing. There are two purposes for setting up an IRB. First of all, these results are used in a continuous reasoning process. Secondly, when a design fails, a backtrack (redesign) is performed using the information stored in the IRB.

18. **Final results base (FRB):** This stores all acceptable schemes in the design. The decision-making module refines the best one from among those stored in the FRB.

19. **Method base:** This records the structural descriptions of all the problem-solving methods. Each parameter or structural attribute has its own specific methods of generation through reasoning, table look-up, analogy, calculation, and so forth. The method base is operated in many different ways. Separated from the problem solver, it partially describes the problem-solving procedure. When handling a new parameter, users may add a new description to the method base without changing the problem solver. All methods are described by object-oriented frames.

20. **Rule base:** This consists of many files organized as production rules. Each file is generated by the knowledge acquisition module to perform a specific sub-task.

21. **Model base:** This stores various parametric models needed in parts assembly. The parameters consist of space position (x, y, x) and orientation (v_x, v_y, v_z). A mechanical model (part or component) can be placed on a proper position if these key parameters and a transformation matrix are provided.

22. **ORACLE™/RDB™:** This is a commercial database under UNIX™/VMS™ operating systems and functions as a center for exchanging information. The databases are the bridge between product design and manufacturing in a CIM environment.

23. **Optimization program base (OPB):** This provides seven optimization techniques, such as linear optimization, nonlinear constraint optimization, discrete optimization, etc. The optimization programs provide the common optimization algorithms. An optimal model of mechanical design described in the method base can automatically link the programs and generate an executable file. The results obtained by running the file are stored in the IRB.

24. **Shortcut estimation module:** The scheme quotation is an important part of product design and is also the foundation of the scheme evaluation and decision-making. Economic analysis, shortcut and estimation of production cost, marketing situation, material price and

profit rate before manufacturing have a practical significance for direct plant production.

25. **Knowledge acquisition module:** This operates the knowledge base that consists of many files. Each file is a collection of rules for special problems. The knowledge acquisition module can maintain all files, for example by skimming, adding, deleting, or modifying rules in one file.

26. **On-line help module:** This brings up a context-sensitive on-line help window that explains system commands. This window can be selected at any time from any module.

4.5 Function and structure conceptual design

4.5.1 Functional concept design

Functional concept (Definition 3) design, as the first stage reasoning, aims at expanding user's specifications into design conditions and restraints. In general, a user may fully describe the specifications and the working environments, but it is difficult to translate these specifications into design conditions and restraints. That is, a user can never know whether his/her information is correct and complete. For example, while designing a chemical plant, where the plant is located has an impact on conceptual design because the utilities available on site, for instance cooling-water temperatures, depend on the geographical location. Similarly, the costs of raw materials reflect the transportation costs, depending on where these materials are produced. The functional concept design makes functional descriptions more specific and supplies some missing information.

Functional concept design needs to use shallow knowledge. In IDIDE, shallow knowledge is defined as simple cause/effect knowledge (Kapp, 1989), which is represented as heuristic rules without mathematical formula or computation in the context. All the conclusions that satisfy the restraints should be triggered, and all effective concepts should be codified in the fact base.

4.5.2 Structural concept design

Structural concept (Definition 5) design (second stage reasoning) is the main content in IDIDE. The module can extract effective structural concepts from the structural concept space. Each concept is a specific symbol corresponding to a structural type, a structural concept, or a structural alternative. Second-stage problem-solving can be divided into five steps: (1) establishment of structural concept space; (2) establishment of the reasoning network; (3) building knowledge base; (4) restraint reasoning; (5) scheme-resolving (or scheme decomposition).

(a) Structural concept space

The structural concept space consists of all possible alternatives of parts or components and structural types. The space can be described by an AND/OR tree. The tree root represents a product or system to be designed and the leaves (or nodes) stand for its components and possible structural alternatives. The arcs (branches) of the tree represent the specific hierarchical relationships among concepts.

In this concept representation network, if there exists an AND relationship between a node and all its subnodes, the node can be implemented only if the subnodes are triggered (or performed). On the other hand, if there exists an OR relationship between a node and all its subnodes, the node can be carried out if only one of its subnodes is triggered. For example, Figure 4.5 shows part of the structural concept space in a wheel loader design.

A typical wheel loader consists of several subsystems such as hydraulic subsystem, transmission subsystem, brake subsystem, working device, etc. Here, if we only consider designing transmission, the corresponding subsystem is broken down into four independent parts in terms of transforming ways: hydraulic, fluid, mechanical and electric transmission. When designing fluid transmission, the tasks performed include the selection of engines, and designs of clutch coupling, torque converter and gear box. A structural concept space (or a concept tree) may consist of hundreds and thousands of concepts or nodes. This tree structure depends on the design thinking of experts; that is, each domain expert has his/her own structural concept tree. Therefore, when building a structural concept space, an intelligence engineer should fully consider different solutions from various experts, and coordinate the conflicts among them.

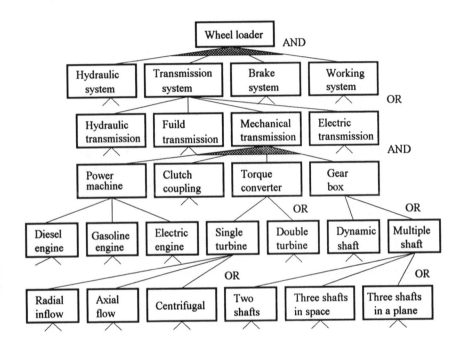

Figure 4.5 Part of a wheel loader structure concept space.

(b) Concept reasoning network

When all functional concepts, specifications and restraints (they are always viewed as the premises of rules) are linked to each structural concept in the structural concept space, this space (or tree) is translated into a concept reasoning network. Since a structural concept may be a premise of a concept node at another branch, the reasoning network becomes an 'AND/OR' graph. Without losing generality, we define that C_i

represents a structural concept and E_i expresses a restraint condition (or a premise). Figure 4.6 (a) shows a simplified structural concept space from Figure 4.5, and Figure 4.6 (b) demonstrates a concept reasoning network that includes a few restraints and functions.

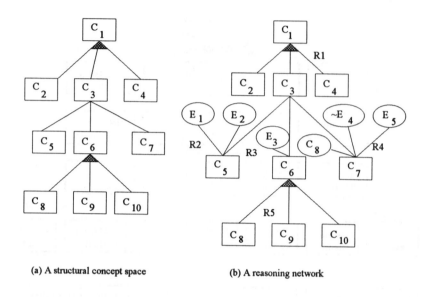

(a) A structural concept space   (b) A reasoning network

Figure 4.6 A simplified structural concept space and reasoning network.

(c) Rule base

The rule base of structure concept design consists of three aspects: production rules, the relationship list between rules and nodes, and the record list of goal nodes.

Rule: In IDIDE, rules are defined as follows:

$\langle Rule \rangle ::= (IF \{ \langle Premise \rangle \}^+ \text{ THEN } \{ \langle Conclusion \rangle \}^+)$
$\langle Premise \rangle ::= (\{ \langle Function \rangle^* \} \{ \langle Element \rangle \}^+)$
$\langle Function \rangle ::= AND, OR, NOT, +, -, \dots, \text{ any predicate or operator}$
$\langle Element \rangle ::= \langle Vector\ Element \rangle \mid \langle Attribute\ Value \rangle$

<Value>:: = <Symbol> | <Number>
<Conclusion>:: = ({<Statement>}$^+$ | {<Action>}$^+$)
<Statement>:: = (<Vector Element>) | <Attribute Value>)
<Action>:: = {<Function>}$^+$ | {<Operator>}$^+$

where

{}* stands for optional,
{}$^+$ stands for one occurrence at least, and
| stands for "or".

The structure concept reasoning network as shown in Figure 4.6 (b) can be converted into the following rules:

(Rules

(R_1 (IF C_1 THEN C_2 C_3 C_4))
(R_2 (IF (AND E_1 E_2 C_3) THEN C_5))
(R_3 (IF (AND E_3 C_3) THEN C_6))
(R_4 (IF (OR (AND C_3 C_8) (AND (NOT E_4) E_5 C_3) THEN C_7))
(R_5 (IF C_6 THEN C_8 C_9 C_{10})))

Relationship list between rules and nodes: In general, the velocity of searching and the efficiency of an inference engine are relatively low since production rules are not a kind of structural knowledge representation. In a production system (or rule-based system), the logical relationship between rules and nodes is implied in the context. In order to improve the reasoning efficiency, IDIDE applies an organization of tree node levels to explicitly stand for the logical relationship. In terms of the relationship of rules, the knowledge acquisition module can automatically classify the rules linking a node with its *forward rules* and *backward rules*, (i.e. each node has its forward rules and/or backward rules, while the root node only has backward rule and the leaf nodes have only forward rules). Here, we define the rules that verify the node (it can be viewed as a conclusion) are forward rules of the node, and the rules that can be triggered by the node (it can be viewed as a premise) are backward rules of the node. This classification can bring out the following three advantages:

- Increasing the reasoning efficiency: The logical relationship has been explicitly represented and automatically classified by the knowledge acquisition module before the system runs. Thus, when the inference engine verifies a node, it can quickly find out the rules concerning the node. A special testing shows that the efficiency can be raised 8 to 10 times after the logical relationship is recorded. Of course, the processing method takes more memory space.

- Improving the explanation ability to reasoning paths: the inference engine can backtrack a reasoning process using the logic relationship.

- Enhancing the function for checking the contradictoriness and redundancy among rules.

Here, we define

$$G = (C, R, Q) \qquad\qquad (4.7)$$

as a concept reasoning network graph,

where

$C = \{c_i\}$ ($i = 1, 2, ... I$) is a concept node set,

$R = \{r_j\}$ ($j = 1, 2, ... J$) is a rule set, and

$Q = \{$After, Before$\}$ represents the relationship between rules and nodes (forward rules and backward rules).

The rules (from R_1 to R_5) discussed above only represent the context relations between concept nodes, but do not give us the logical relationship between nodes and rules (i.e. nodes and arcs). To clearly illustrate this relationship in correspondence to Figure 4.6 (b), we set up a *Before* list and an *After* list. The Before list records all forward rules and related nodes, while the After list stores all backward rules and related nodes.

(After (C_1 ($R_1 R_2 R_3 R_4 R_5$)

  C_3 ($R_2 R_3 R_4 R_5$)

  C_6 (R_5)))

(Before (C_8 ($R_1 R_3 R_5$)

$C_9 (R_1 R_3 R_5)$
$C_{10} (R_1 R_3 R_5)$
$C_5 (R_1 R_2)$
$C_6 (R_1 R_3)$
$C_7 (C_1 R_4)$
$C_2 (R_1)$
$C_3 (R_1)$
$C_4 (R_1)))$

Record of goal nodes: A concept reasoning network is equivalent to a concept set required by domain experts to solve their problems. The operation of the reasoning network should not change the existing logical relationship between concepts. Therefore, a record list about goal nodes, namely *goal list*, is included in the rule base. This list can control the depth of searching tree. For example, the goal list for Figure 4.6 (b) is expressed as

$$(GOAL (C_2 \ C_4 \ C_5 \ C_7 \ C_8 \ C_9 \ C_{10}))$$

If the goal list is not set up and E_3 and C_1 are true, the rules of R_1, R_3, R_4, and R_5 are triggered. In this case, the reasoning result is stored as an effective concept space as shown in Figure 4.7 (a), where C_7 obviously has lost its original logical relationship with C_3 because C_8 is not only a *goal node*, but also a premise of C_7 (Comparing with Figure 4.6 (a)). However, if a goal list is used, the original logical relationship can be maintained (see Figure 4.7 (b)).

(d) Restraint reasoning

Conventional reasoning methods can obtain conclusions based on some assumptions. Such an inference engine usually employs *forward chaining* control strategy (*data-driven*) or *backward chaining* (*goal-driven*). The objective of *restraint reasoning* is to convert a structural concept space into an effective concept space that satisfies design restraint conditions. In other words, this inference can prune the original structural concept space based on the facts provided by the users. This restraint reasoning engine is used in IDIDE.

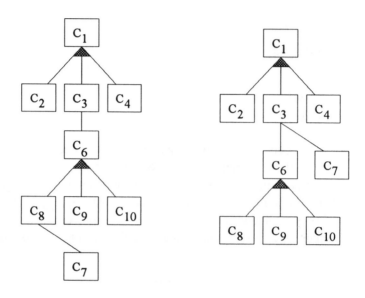

(a) C7 losing its logical relationship   (b) C7 keeping its logical relationship

Figure 4.7 Functions of goal list.

In restraint reasoning, the verification of a single concept node (a conclusion or an assumption) is not enough. Instead, all nodes in the structural concept space should be verified. An important criterion to identify unsuccessful reasoning is to find the situation when an 'AND' node is not triggered or none of the 'OR' nodes is activated. Therefore, no design scheme or pattern is produced. Restraint reasoning employs a bidirectional control strategy (combination of forward chaining and backward chaining). That is, the subnodes are expanded by using forward chaining reasoning to find backward rules for these nodes, then, push them into a *stack*. Backward chaining reasoning can verify the expanded concept nodes stored in the stack.

To introduce the *restraint reasoning algorithm,* some tokens used in the algorithm are explained as follows:

Stack: to record the expanded nodes,
Static Facts: to store the facts associated with a design,
Intermediate Results: to record intermediate results and asked facts,
Final Results: to record the final effective concepts.

Restraint reasoning algorithm:

(1) Give C_i (In general, C_i is a design goal).

(2) Put C_i in Stack, Static Facts and Final Results.

(3) Check whether Stack is empty. If empty, show the effective concept space. Then, exit.

(4) Take C_i from Stack, then, find the backward reasoning rules of C_i and all the forward nodes associated with the rules (i.e. all subnodes of C_i).

(5) If C_i has a subnode (i.e. C_i is not in the goal list),

   (i) if the subnode is not a goal node, put C_i in Stack and go to (3). Otherwise,

   (ii) the subnode is a goal node, continue.

(6) If C_i has no subnodes, or C_i is a goal node, verify the premises of forward rules associated with C_i, then

   (i) try to match C_i with the restraints in Static Facts, Intermediate Results and Final Results and record the rules triggered, then, go to (3). Otherwise,

   (ii) consult a user, and push the facts obtained from him/her into Intermediate Results. If C_i is verified, put C_i in Final Results and record the rules triggered, then, go to (3). Otherwise,

   (iii) if C_i is not verified, put the negated conclusion ($\sim C_i$) in Intermediate Results, give up C_i, then, go to (3).

Figure 4.8 is an example to illustrate the symbolic operation based on Figure 4.6 (b), and the reasoning result is shown in Figure 4.7 (b).

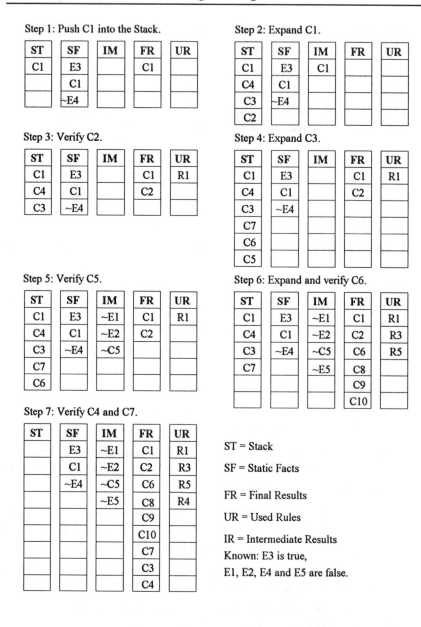

Step 1: Push C1 into the Stack.

ST	SF	IM	FR	UR
C1	E3		C1	
	C1			
	~E4			

Step 2: Expand C1.

ST	SF	IM	FR	UR
C1	E3	C1		
C4	C1			
C3	~E4			
C2				

Step 3: Verify C2.

ST	SF	IM	FR	UR
C1	E3		C1	R1
C4	C1		C2	
C3	~E4			

Step 4: Expand C3.

ST	SF	IM	FR	UR
C1	E3		C1	R1
C4	C1		C2	
C3	~E4			
C7				
C6				
C5				

Step 5: Verify C5.

ST	SF	IM	FR	UR
C1	E3	~E1	C1	R1
C4	C1	~E2	C2	
C3	~E4	~C5		
C7				
C6				

Step 6: Expand and verify C6.

ST	SF	IM	FR	UR
C1	E3	~E1	C1	R1
C4	C1	~E2	C2	R3
C3	~E4	~C5	C6	R5
C7		~E5	C8	
			C9	
			C10	

Step 7: Verify C4 and C7.

ST	SF	IM	FR	UR
	E3	~E1	C1	R1
	C1	~E2	C2	R3
	~E4	~C5	C6	R5
		~E5	C8	R4
			C9	
			C10	
			C7	
			C3	
			C4	

ST = Stack

SF = Static Facts

FR = Final Results

UR = Used Rules

IR = Intermediate Results

Known: E3 is true,

E1, E2, E4 and E5 are false.

Figure 4.8 Example to illustrate the restraint reasoning strategy.

(e) Scheme-resolving (or separation)

Generally, an effective structural concept space is an assemblage of a number of design schemes. For example, Figure 4.7 (b) can be divided into two acceptable schemes as shown in Figure 4.9. The number of acceptable schemes in an effective concept space is given as:

$$NM = \prod_{i=1}^{n} t_i \qquad (4.8)$$

where

NM = the number of the acceptable schemes,

n = the number of 'OR' nodes on an effective concept tree, and

t = the number of 'OR' node branches.

Obviously, there exists a vast amount of schemes as n and t increase. However, due to the use of domain expertise knowledge and heuristics, the effective concept space can be confined.

Scheme-resolving divides an effective concept space (a tree) into several subtrees containing only 'AND' nodes and a single 'OR' node. Each subtree can be an acceptable scheme. For example, the effective concept space as shown in Figure 4.7 (b) can be expressed as the following list:

(Schemes $(C_1 (C_2 C_3 (C_6 (C_8 C_9 C_{10}) C_7) C_4)))$.

After the resolving operation, this list is separated into two subtrees (or sublists) as shown in Figure 4.9:

(Scheme 1 $(C_1 (C_2 C_3 (C_6 (C_8 C_9 C_{10})) C_4)))$, and

(Scheme 2 $(C_1 (C_2 C_3 (C_7) C_4)))$

Let us now discuss the algorithm for scheme-resolving. An effective structure concept space is defined as:

Schemes = (EC, UR, Q) \qquad (4.9)

where

$EC = \{ec_i\}$ $(i = 1, 2, ... I)$ represents a set of effective concept nodes,

$UR = \{ur_j\}(j = 1, 2, \ldots J)$ represents a set of triggered rules, and

$Q = \{After, Before\}$ represents the logical relationship between effective nodes and triggered rules.

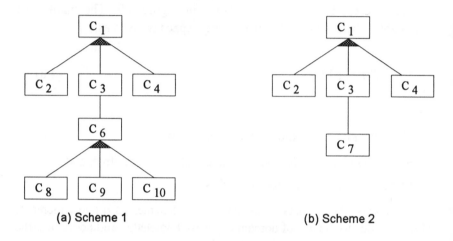

(a) Scheme 1   (b) Scheme 2

Figure 4.9 Scheme-resolving.

Scheme-resolving algorithm: (*Scheme* records all acceptable schemes.)

(1) Push Final Results data into Stack.

(2) Verify whether Stack is null, if yes, separately display all acceptable schemes in Scheme, then exit.

(3) take a list L_i from Stack, and search an 'OR' node in L_i,

    (i) If the 'OR' node is found, and it has more than two branches, L_i should be divided into sublists from this node. Push these sublists into Stack, and go to (2).

    (ii) If no 'OR' node is found, put L_i in Scheme, and go to (2).

4.6 Parameter design

The parameter design determines the attributes and attribute values of structure types (or structure concepts). The structural parameter design accomplishes the detailed descriptions of system or product structures, and provides necessary data and information for quantitative analysis and qualitative evaluation. Parameter design is not a single symbolic reasoning problem. It also deals with numerical computation, table look-up and curve observation. Therefore, its knowledge representation should not express only problem-solving methods but also the description of problem-solving processes. It uses several problem-solving strategies, and deals with solution conflicts between different design methods. Since many design methods can be chosen to solve the same problem in engineering design, a better design often depends on design requirements, application environment, as well as experts' private knowledge. The development of knowledge base, data structure and control strategy of the parameter design module takes the above considerations into account.

4.6.1 Knowledge base structure of parameter design

The knowledge used in parameter design is deep knowledge (model-based knowledge) (Kapp, 1989). In IDIDE, an object-oriented data structure (or frame) is employed to describe the knowledge. The objects (frame names) usually represent structural concepts, structure types or equipment tokens. The assemblage of all frames is stored in the method base, which can be executed with the problem solver. Since the method base is independent of the problem solver, it can be easily updated without adjusting the problem solver.

4.6.2 Frame structure

Frame is a special data structure to represent domain knowledge. One of its advantages is the capacity to describe the attribute values of an entity (through slots and facets of a frame) and the logical relationships among entities (for example, using *is_a* and *has_a* slots). IDIDE here makes use of frames to describe problem-solving methods and problem-solving processes.

A frame comprises a frame name, several slots and values. Slots represent attributes of an object, values express attribute descriptions and a frame name stands for a specific object. In IDIDE, three formulated frames are developed to define problem-solving strategies: symbolic reasoning frame, numerical computation frame and table look-up or curve observation frame. These frames are structured in the following ways:

Numerical computation frame:

Frame name: (string) x x x

(1) Knowledge source:(string) x x x

(2) Setting date: (string) x x x

(3) Task level: (string) x x x

(4) Task content: (list) x x x

(5) Solution specification: (list) x x x

(6) Search path: (list) x x x

(7) Inquiry agency: (list) x x x

(8) Computation formula:(list) x x x

(9) Relation frame: (list) x x x

(10) Explanation: (string) x x x

(11) Default method: (list) x x x

Symbolic reasoning frame:

Needs to add two slots to the above frame description:

(12) Rule set: (string) x x x

(13) Inference engine type: (string) x x x

Table look-up frame:

For table look-up frame, data table and interpolation method slots must be given:

(14) Data table: (string) x x x

(15) Interpolation method: (string) x x x

The three frames above are already defined in IDIDE. If a new problem-solving method is required, a new frame structure should be defined. The functions of all slots are briefly explained below:

(1) Knowledge source: The slot records the source of knowledge.

(2) Setting date: This describes when the frame or the problem-solving method was set up to make it easier for users to modify and update the method base.

(3) Task level: This represents the task hierarchy; i.e. the relationship of structural concepts.

(4) Task content: This describes the content of the problem to be solved, such as what parameters are used to describe the current problem.

(5) Solution specification: This specifies the parameters related to the current problem to be solved. The parameters in the slot are the premises; that is, the conditions to refer the formula in the computation formula slot.

(6) Search path: This illustrates the depth of searching Static Facts, Dynamic Facts, Intermediate Results and Final Results.

(7) Inquiry agency: This inquires users about information when some parameters cannot be found in the fact bases and databases.

(8) Computation formula: This slot stores the analytical formula or numerical computation models.

(9) Relation frame: This describes the logical relationship between the current frame (or structure type) and other related frames.

(10) Explanation: This records the descriptions about variables, concepts and problem-solving strategies. The textual information can be directly shown on screen if required.

(11) Default method: This slot can provide another problem-solving method or default solutions when a chosen method in the computation formula slot fails to solve the problem.

(12) Rule set: The slot records the file name of a rule set, which may be used in symbolic reasoning.

(13) Inference method: This stores the file names of inference engines. Since different problems require different reasoning strategies, IDIDE provides four reasoning strategies.

(14) Data table: This records the names of data tables in a file or a database.

(15) Interpolation method: This stores a file name of interpolation methods that may be used in problem-solving.

The analogy is a usable design methodology, which is often employed in conceptual design or design synthesis. IDIDE provides such a function. Users can obtain the statistical formula (models) from previous design and manufacturing data of existing products, and put the formula in the default method slot. If the problem solutions cannot be obtained from the mathematical models, IDIDE uses the statistical models to get an approximate solution.

4.6.3 Method base

The method base consists of all frames with each frame corresponding to a structural type, parameter, or an empirical coefficient. The frames are interrelated frames, and in IDIDE this relationship is simply treated as a kind of level relation or dendritic relation. Therefore, one frame only depends on the other vertically.

The levels of frames are automatically generated by the scheme solving. For instance, two acceptable schemes are generated for Figure 4.9. In scheme 1, there are eight frames on four levels on which the problem-solving relies. The problem solver can manipulate the frames from bottom to top.

4.6.4 Problem solver

The problem solver identifies the function and structure of a frame, and then processes the slots of the frame in order of frame levels. Figure 4.10 shows a block diagram of the problem solver. The problem-solving strategy provides the following special functions:

(1) Coupling symbolic reasoning with numerical computation: Symbolic reasoning, numerical computation and table look-up are used to solve a problem.

(2) Executing the default method slot: If a specified method fails, IDIDE can execute the default method slot to obtain a statistical value or an approximate solution.

(3) Solution provided by an expert: If all above problem solving strategies fail, IDIDE can require an expert to provide solutions.

(4) Asking for expert's suggestions: If an expert disagrees with the solutions generated by IDIDE, the expert has the priority to modify the solutions.

4.7 Scheme analysis

Parameter design provides the data and information for analyzing the product or system performances. Since the analysis contents and methods vary with products, IDIDE cannot include all analysis programs for general applications. However, its internal interface can link the existing analysis models, commercial computation tools and packages as needed. It can implement data transformation and communication between IDIDE and analysis programs. In addition, the internal interface can call packages written in different languages such as C™, Pascal™ and FORTRAN™.

During analysis, all generated schemes should be examined, and the analysis results stored. Some schemes are eliminated if they cannot satisfy design requirements and constraints.

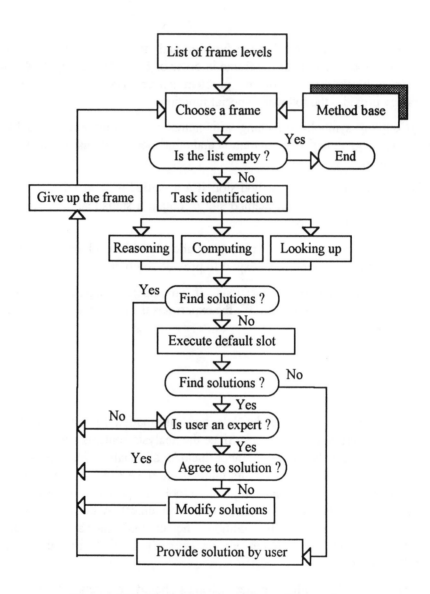

Figure 4.10 Block diagram of the problem solver.

4.8 Comprehensive evaluation

The comprehensive evaluation module determines whether the results (or schemes) of conceptual design are feasible or practicable, and evaluates the quality of a design scheme. In addition, it also provides the detailed feedback information for improvement of design schemes (redesign) as well as an optimal backtrack point (or position) to avoid blind search in a new design and reduce iterations. IDIDE implements a general evaluation algorithm suitable for estimating the comprehensive performance of a mechanical product or system in the early conceptual design stage. For a given design scheme, if the users gives an index (target) system and the weight of each index, the evaluation module in IDIDE can perform a comprehensive estimation based on the index system and index weights.

4.8.1 Control structure

Figure 4.11 shows a control structure for the evaluation module and feedback redesign module. This control configuration provides the following functions:

- to generate a final conclusion that presents the comprehensive performance of a design scheme using a hierarchical index system;

- to simulate the performances of the schemes using analysis results;

- to arrange the acceptable schemes in order of the design quality with a multi-hierarchical fuzzy evaluation method;

- to provide the feedback information and an optimal backtrack point for the redesign model.

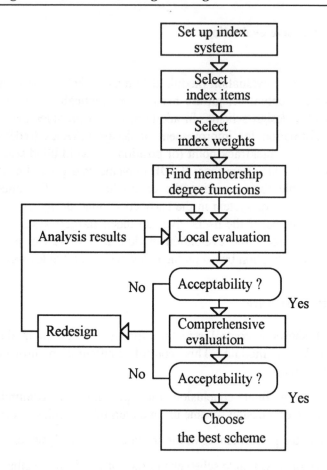

Figure 4.11 Control structure for comprehensive evaluation and redesign.

4.8.2 Comprehensive evaluation indices

An index system is a systematized and formalized description of performances of a product or system to be designed, and reflects the integrated performance of the design object and the relationship between subsystems. A multi-hierarchical model (Chen, 1983) is adopted in IDIDE, which uses a tree structure to represent the evaluation targets horizontally, and their sub-targets vertically (Figure 4.12). In general, the row number of targets (tree depth) should be less than 4, and the column number (tree width) less than 10. This reflects the fact that if an index system is too big, errors from various circumstances will impair the evaluation quality. For a specified mechanical product or system, it is very important to set up the index system and select index weights.

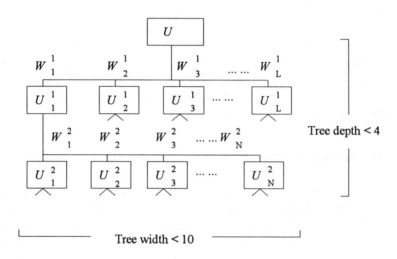

Figure 4.12 Tree structure of the index system.

Items of the index system represent the detailed descriptions of specifications. They reflect the performances of a product to be designed objectively and comprehensively. Therefore, a variety of factors with impact on the performance should be fully considered. Obtaining the global optimum is always a fundamental rule when selecting the index system. A mechanical product is a complicated system in which exist a variety of interdependent and interactive relationships between subsystems. As a result, the quality of a product depends not only on the quality of its subsystem, but also on the integration quality of the product components. Since the index items reflect the comprehensive performance, the system acceptability relies on the entire design optimization.

For a specific index system, its top items are abstract descriptions of characteristics of a whole mechanical product, and its bottom items are specific descriptions. The bottom items are directly associated with the design parameters generated by the analysis module. When a design fails, the redesign module improves the design scheme by modifying the values of these items.

4.8.3 Index weights

The index weight system is also a tree structure as shown in Figure 4.12. IDIDE uses two methods to select index weights. One is mathematical statistics (Zhou, *et al.*, 1989a). This method sometimes fails because of the difficulties in collecting statistic data. Another is an empirical method to get the index weights directly from domain experts.

4.8.4 Membership degree function

After the index values are assigned, a designer checks whether the design is 'satisfactory' or 'unsatisfactory'. The degree of 'satisfactory', for a new product design usually remains problematic. Therefore, the membership degree concept in fuzzy theory is introduced into the evaluation module to make a quantitative evaluation on various levels of a mechanical system.

Suppose a possible value for a given index is $R \in [a, b]$, and a satisfactory value for the index is represented by A, then μ_A is a membership degree function of concept A for the index value, as described in Equation 4.10.

$$\mu_A = f(R) \qquad R \in [a, b] \qquad (4.10)$$

Many methods to determine membership degree functions have been developed (He, 1983). IDIDE uses an expert-evaluation method (Zhou *et al.*, 1989a).

4.8.5 Comprehensive evaluation algorithm

Comprehensive evaluation is the qualitative estimation of the quality of a single design scheme. The evaluation is performed step by step from the lowest level up to the highest in an index system. Finally, a summary (or conclusion) results from the evaluation of the first level (or the tree root) in the index system.

In the evaluation at each level, feasibility is considered first. That is, the module can find only whether a scheme satisfies all customers' requirements and design specifications on the basis of analysis results (index values) and membership functions. The evaluation is divided into two stages. Local evaluation (e.g. estimating individual aspects of the comprehensive performance) is executed first. Then, a comprehensive evaluation on the design scheme is performed. The module finally provides the following information:

- comprehensive evaluation values,
- feasible indices and weights, and
- satisfaction degree of indices.

To better understand the comprehensive evaluation algorithm, a simple example of third level evaluation is given below:

Suppose the index vector for the third level (Figure 4.12) is

$$U_3 = (U_1^3, U_2^3, ..., U_L^3).$$ (4.11)

The corresponding value for each index is

$$R_3^i \in [r_{i1}, r_{i5}] \quad i = 1, 2, ..., L.$$ (4.12)

The corresponding membership degree function for each index is

$$\mu_i(r), \quad r \in R_3^i \quad i = 1, 2, ..., L.$$ (4.13)

The index weight vector (Figure 4.12) is

$$W_3 = (W_1^3, W_2^3, W_3^3, ..., W_L^3),$$ (4.14)

and the index values computed by analysis model are

$$P_3 = (P_1^3, P_2^3, P_3^3, ..., P_L^3).$$ (4.15)

Algorithm for third level evaluation:

(1) Calculate $U_i^3(P_i^3)$, $i = 1, 2, ..., L$.

(2) If $U_i^3(P_i^3) \geq U_{min}^3$ $i = 1, 2, ..., L$,

   then calculate $C_3^L = \sum_{i=1}^{n} W_i U_i(P_i^3)$.

   Otherwise, fill up the table that records error messages.

(3) If $C_3^L \geq M_2^L$ (where, M_2^L = threshold value of the higher level),

   then output the result of calculation.

   Otherwise, fill up the error table.

The general evaluation procedures for other levels are similar to the above one for the third level.

4.9 Decision-making

To order all acceptable schemes in terms of design quality, IDIDE uses the comprehensive decision method based on multi-factor multi-level decision-making models (He, 1983). If N acceptable schemes are generated in conceptual design, the decision-making module can order the schemes and produce a quantitative evaluation solution.

The multi-factor multi-level decision-making approach can order all acceptable schemes by synthesizing multiple evaluation indices and their weights, and quantitatively evaluate the comprehensive performances of all design schemes. As an example, let us discuss the decision-making model of a signal-level index system.

Suppose the new design is of number N of evaluation indices, the vector μ stands for the factor field of an index system as described in Equation 4.16.

$$\mu = (\mu_1, \mu_2, \ldots, \mu_N) \tag{4.16}$$

If a number N of acceptable schemes are generated in product design, the decision-making matrix $\tilde{R}$ is expressed as:

$$\tilde{R} = \begin{vmatrix} r_{11} & r_{12} & \cdots & r_{1M} \\ r_{21} & r_{22} & \cdots & r_{2M} \\ \vdots & \vdots & \cdots & \vdots \\ r_{N1} & r_{N2} & \cdots & r_{NM} \end{vmatrix}. \tag{4.17}$$

where r_{ij} are the membership degree functions of the ith design scheme and the jth evaluation index.

If the index weight vector is given as below:

$$W = (w_1, w_2, \cdots, w_N), \tag{4.18}$$

the comprehensive decision-making vector $\tilde{B}$ is computed by equation 4.19 as follows:

$$\tilde{B} = W \oplus R = (b_1, b_2, \cdots, b_N). \tag{4.19}$$

where

$$b_j = \sum_{i=1}^{N} w_i r_{ij},$$

w_i and r_{ij} are real variables,

$\displaystyle\sum_{i=1}^{N}$ represents a sum operation with the operator '$\oplus$',

the operator '$\oplus$' stands for that $a \oplus b = \min(1, a + b)$.

Because b_j represents the comprehensive evaluation result of the jth design scheme, the best design scheme can be selected from among acceptable schemes based on the values of b_j. This decision-making model, namely Model I, is used in evaluating a signal-level index system. In real design, most of index systems are multiple levels. Therefore, the multi-factor multi-level design-making model in IDIDE is developed for complex index systems. The implementation steps are described below:

Step 1: According to the structure of an index system, all index items are classified, and the index levels are selected. For example, if an evaluation index system is divided into three levels, their factor fields are expressed as:

level 1 $\mu_1^0 = (\mu_1^{01}, \mu_2^{01}, \cdots, \mu_N^{01})$ (4.20)

level 2 $\mu_2^1 = (\mu_1^{21}, \mu_2^{21}, \cdots, \mu_M^{21})$ (4.21)

$\mu_2^2 = (\mu_1^{22}, \mu_2^{22}, \cdots, \mu_M^{22})$

$\cdots$

$\mu_2^M = (\mu_1^{2M}, \mu_2^{2M}, \cdots, \mu_M^{2M})$

level 3 $\mu_3^1 = (\mu_1^{31}, \mu_2^{31}, \cdots, \mu_L^{31})$ (4.22)

$\cdots$

$\mu_3^L = (\mu_1^{3L}, \mu_2^{3L}, \cdots, \mu_L^{3L})$

Step 2: Each μ_k^l is computed using Model I from the bottom up, i.e.

$W_k^l \oplus R_k^l = B_k^l$ (4.23)

where W_k^l, R_k^l are the weight vector and decision-making matrix of μ_k^l respectively.

Step 3: the multi-level decision-making model is performed.

The multi-factor multi-level decision-making model has been used in the wheel loader conceptual design expert system (WLCDES). A comprehensive decision-making result is shown in Table 4.16.

4.10 Redesign

The five stages of the problem-solving strategy can be accomplished in such a way as discussed above. However, a special situation has to be considered in which all design schemes cannot meet customers' requirements and specifications, i.e. no acceptable scheme is generated. In this case, redesign is required. This uses the feedback information from the analysis and evaluation modules. In IDIDE, a mathematical model of redesign has been developed.

This redesign method first divides the integrated performance of a mechanical product or system into several levels of indices. The corresponding membership degree of each index serves as a decision variable. Then a computation model is constructed using these variables. IDIDE can link the model with a program of constrained nonlinear optimization in the OPB to generate an executable redesign program. The new index values, i.e. the output after running the program are the results of redesign.

Figure 4.13 shows how an optimal feedback mathematical model is established. M represents goal functions, s.t. is constraint conditions, and Opt. stands for optimal solutions. The following is an example that illustrates how to set up the optimal model on the second level in the case of index M.

Suppose that the evaluation index vector and weight vector of M_{12} are denoted as U_{12} and W_{12}, respectively

$$U_{12} = (U_{121}, U_{122}) \quad \text{and} \quad W_{12} = (W_{121}, W_{122}). \tag{4.24}$$

The corresponding membership degree vector of evaluation index vector U_{12} is

$$D_{12} = (W_{121}d_{121}, W_{122}d_{122}), \tag{4.25}$$

and the ideal membership degree (which corresponds to the best index value) is

$$\overline{D}_{12} = (W_{121} \overline{d}_{121}, W_{122} \overline{d}_{122}). \tag{4.26}$$

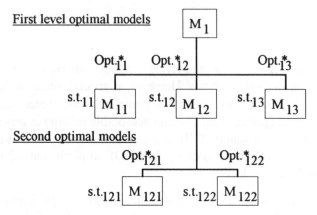

Figure 4.13 An example of setting up optimal mathematical models.

An optimal model for redesign at the second level is

$$\text{Min} \left\| D_{12} - \overline{D}_{12} \right\|$$

$$\text{s.t.} \quad \sum_{i=1}^{2} W_{12i} \, d_{12i} \geq d_{12i}^{*}$$

$$g_{12i}^{\text{max}} \geq d_{12i} \geq g_{12i}^{\text{min}} \quad (i = 1, 2)$$

$$(4.27)$$

where, g_{12i}^{max} and g_{12i}^{min} are the upper and lower limits for membership degree, and d_{12i}^{*} is the distributed value of index M_{12} from the optimal models at the first level (see Figure 4.13). If the solution is

$$D_{12}^{*} = (W_{121} \, d_{121}^{*}, W_{122} \, d_{122}^{*}),$$

$$(4.28)$$

then D_{12}^{*} is the optimum distribution of M_{12} indices. It also expresses the comprehensive performance of the third level indices M_{121} and M_{122}.

4.11 Applications

Three application systems for conceptual design of mechanical products using IDIDE have been developed. They are:

- Conceptual design expert system for transmission (CDEST);
- Wheel loader conceptual design expert system (WLCDES); and
- Turbine design expert system (TDES).

4.11.1 CDEST

(a) Functions of CDEST system

CDEST is an integrated distributed intelligent system for conceptual design of mechanical transmission, which combines expert systems with conventional CAD techniques, and can be used to design the general scheme of transmissions according to customer's requirements.

CDEST can implement various designs (including power utility, shift, layout sketch), a variety of selections (such as bearing, lubricating, connecting, sealing), as well as part parameters designs (for example, property parameters, function parameters, geometry parameters, position and orientation parameters). In addition, it can also carry out rigidity and strength stress analyses for important parts, evaluation of final results (or design schemes), and shortcut estimation of manufacturing costs. The information and data (numerical and symbolic) generated lays a foundation for the follow-on detailed design of parts, thus creating the desirable conditions for an integrated CAD system. Another function of the system is to generate the rough assembly drawings and complete the layout space design. It allows the users to modify the drawings and to influence the system's decision-making.

To better maintain the knowledge base and easily add new knowledge, CDEST provides a special knowledge acquisition module which can be used to modify and expand the knowledge base. According to engineering design requirements, the system can carry out conceptual design, layout design and shortcut estimation for various types of transmission devices.

(b) Design principles of CDEST

The key issues in developing CDEST are how to couple expert systems with conventional CAD techniques, and how to meet the specifications under a CIM environment. The basic principles employed to build CDEST are as follows:

- The research and development activity must satisfy the software engineering requirements for CIM.

- Two or three domain experts are designated as the people to extract domain specific knowledge from.

- The system must ensure the segregation of the knowledge base from the inference engines to allow the users to reorganize different knowledge models and domain expertise efficiently.

- To efficiently maintain the knowledge base, the system should include a knowledge acquisition subsystem.

- CommonLisp™ is used to implement the system, and process symbolic information. Numerical information will be processed by FORTRAN™, PASCAL™ or C™.

- The system can exchange information with ORACLE™ database management system.

- A user-friendly environment is required.

- When solving problems, the system can interpret solutions clearly for users.

- The system should contains various problem-solving strategies.

- The system has to be evaluated through real industrial product designs.

(c) Illustration

CDEST is implemented on a SUN SPARCstation 1+™, and provides a heterogeneous knowledge-integration environment. Its symbolic reasoning system is developed in CommonLisp™, while geometric conceptual design and simulation system are programmed in C™ language. FORTRAN™ is used to develop the evaluation module, optimization package and analysis module. A graphics package (CGI™) is run at a SUN™ graphics workstation.

Table 4.1 Specifications of design

Input parameters	Values
Tab-Surface-W (mm):	1000
Tab-Surface-W (mm):	3000
Space-Width (mm):	500
Space-Height (mm):	500
Space-Length (mm):	500
Table-Load(N):	8000
Max.-Feed-V (mm/min.):	1000
Min.-Feed-V (mm/min.):	10
Rapid-Traverse (mm/min.):	0
Spindle-Power (kW):	15
Max.-Spindle-V (r/min.):	630
Min.-Spindle-V (r/min.):	50
Feed-Travel(mm):	0
Move-Part-Weight (kg):	0
Millsprindle-Max-T (N.m):	0
Cutter-Diameter (mm):	0
Mill-Head-Number:	1
FB-Model:	FB001
Plano-Type:	Small
Feed-Box-Type:	Mill-H-F
Driven-Type:	T-Lead-SC
I/O-Shaft-D:	Orthogonal
Mill-Head-D:	Mill-Head-H
Mill-Head-P:	Left

The following tables and figures illustrate a design process of a transmission system. This description discusses the design process step-by-step and presents the design results.

- First of all, a user completes several input pages about design requirements (Table 4.1). After the user confirms them, he/she types a 'DONE' button.

- The structural concept design can determine various structural forms, materials, etc. Table 4.2 shows the specific matrix to present a structure form. The system can automatically generate the matrix, and the user can also modify it.

Table 4.2 Matrix to represent scheme 1

Structure Form: 1[3]-2[4]-3[3]-4[1]						
shaft1	d_gear3	dn_gear2	d_gear1	0	0	0
shaft2	d_gear41	d_gear1	dn_gear1	0	d_gear41	0
shaft3	d_gear4	0	0	0	dn_gear41	d_gear5
shaft4	0	0	0	0	0	dn_gear5

- After structural concept design, the system gives users two design schemes. Figure 4.14 shows their rough layout form. Figure 4.15 is their rotational speed diagram expressing speed change. Figure 4.16 represents a more detailed structure drawing, which offers data and information for analysis and evaluation. Figure 4.17 shows the result of side layout design. Side layout design here means determining position and orientation of shafts.

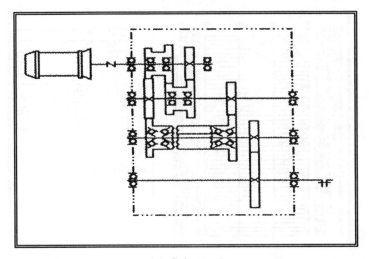

(*a*) *Scheme 1*

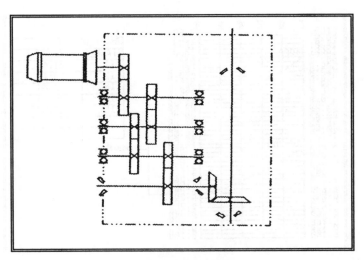

(*b*) *Scheme 2*

Figure 4.14 Sketch of transmission.

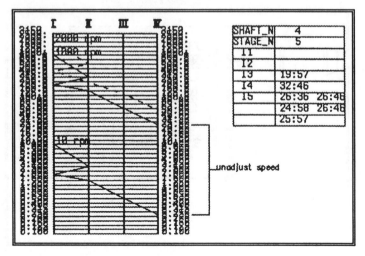

(*a*) *Scheme 1*

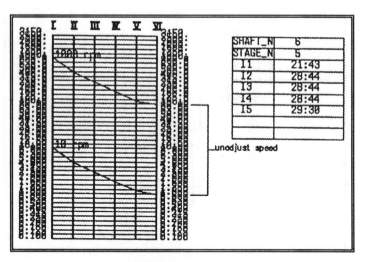

(*b*) *Scheme 2*

Figure 4.15 Rotational speed diagram.

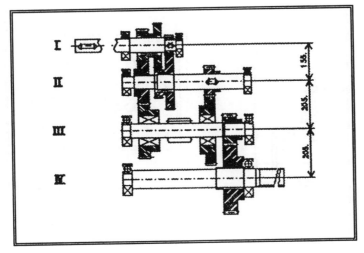

(*a*) *Scheme 1*

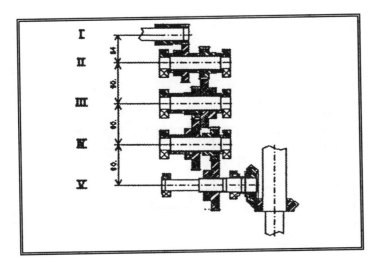

(*b*) *Scheme 2*

Figure 4.16 Structure drawing.

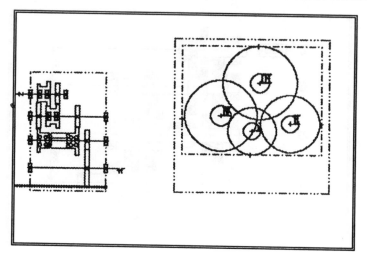

(*a*) *Scheme 1*

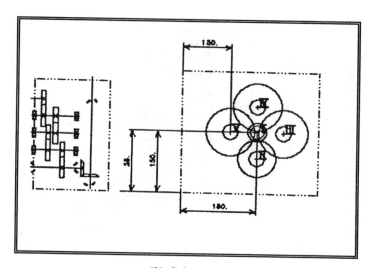

(*b*) *Scheme 2*

Figure 4.17 Side drawing.

- Table 4.3 through Table 4.7 represent the results of parameter design for scheme 1, including number of transmission stages, shafts and gears; parameters of the selected motor; a variety of drive ratios; materials; geometric parameters; machining requirements of shafts, gears; and so on. These results provide important information for scheme analysis and evaluation.

Table 4.3 Number of stages, shafts and gears

Types	Number
Stages	5
Shafts	4
Gears	11

Table 4.4 Parameters of motor

Parameters	Values
Power source type	DC motor
Power	10
Maximum rotation (rpm)	1000
Minimum rotation (rpm)	10

Table 4.5 Drive ratio

Drive ratios	Values
Drive ratio 1	3.0
Drive ratio 2	3.0
Drive ratio 3	18/13
Drive ratio 4	23/18
Drive ratio 5	57/25

Table 4.6 Results of shafts

Shaft	Type	Material	D (mm)	Rotation	Torque
Shaft 1	C-Shaft	45-Steel	45	2000	48.4
Shaft 2	C-Shaft	45-Steel	70	335	285.1
Shaft 3	C-Shaft	45-Steel	80	262	364.3
Shaft 4	C-Shaft	45-Steel	90	115	830.1

Table 4.7 Results of gears

Gear	Shape	Mode l	Num.	Material	Precision	Treatment
Gear 1	SG	4	19	45CR	8-DC	teeth-G48
Gear 2	SG	4	57	45CR	8-DC	teeth-G48
Gear 3	SG	4	32	45CR	8-DC	teeth-G48
Gear 4	SG	4	46	45CR	8-DC	teeth-G48
Gear 5	SG	5	26	45CR	8-DC	teeth-G48
Gear 6	SG	5	36	45CR	8-DC	teeth-G48
Gear 7	SG	5	46	45CR	8-DC	teeth-G48
Gear 8	SG	5	25	45CR	8-DC	teeth-G48
Gear 9	SG	5	25	45CR	8-DC	teeth-G48
Gear 10	SG	5	57	45CR	8-DC	teeth-G48
Gear 11	SG	5	23	45CR	8-DC	teeth-G48

- Table 4.8 presents the conclusions if the results of scheme 1 analysis satisfy the conditions of the working environment. These analyses consist of stress and safety of gears, shafts, bearings and box.

- Table 4.9 shows the results of comprehensive evaluation. Four items are selected as final targets; that is, working efficiency, product volume, manufacturing time and cost. Similarly, we may complete the parameter design, scheme analysis and evaluation of scheme 2. Finally, a better conceptual design can be determined based on the results of comprehensive evaluation.

Table 4.8 Results of gear analysis

Gear	Stress Count (Mpa)	Stress Allow (Mpa)	C-safe Coef. (Mpa)	C-safe Count (Mpa)	Bend-stress allowed	Bend-stress Coef.	Satis-faction
Gear 1	246.3	1370	354.0	22.48	354.0	15.75	OK
... ...	... ...	... ...	... ...	... ...	... ...	... ...	... ...
(omit)	... ...	... ...	... ...	... ...	... ...	... ...	... ...

Table 4.9 Results of evaluation

Type	Efficiency	Volume M^3	M-Time (hours)	Cost ($)
Calculated	59.56%	0.0247	111.91	1892.71
Allowable	50.00%	0.0250	150.00	2500.00

4.11.2 WLCDES

(a) Function of WLCDES

The objective of WLCDES (wheel loader conceptual design expert system) is to provide the required functions, structural concepts and optimal general parameters for the follow-on detailed design of wheel loaders. These parameters include functional, geometric, restraint and positional parameters. All design results are stored in RDB™ (a commercial relational database running on Micro VAX II™), and called by other subsystems in the detailed design stage.

WLCDES is mainly used to design the general schemes of wheel loaders. Its design contents consist of:

• concept design: determining essential structural concepts and types,

- parameter design: providing optimal or near-optimal functional and structural parameters,
- layout design: selecting the existing components and parts, and
- empirical coefficients and detailed specifications: determining the global empirical variables that can be used by subsystems.

WLCDES runs on Micro-VAX II™ computers. Its hardware platform configuration is shown in Table 4.10.

Table 4.10 Hardware platform of WLCDES

Hardware equipment	Model
Host computer	Micro VAX II™
Workstation	GPX or 3100™
Terminal	DEC 200™
Magnetic tape unit	DEC TS05 1600 BPI™
Disk drive	DEC RA81 456MB™
Printer	DEC LP25™

WLCDES makes use of different programming languages and compilers to implement various design tasks. Table 4.11 illustrates the languages and software systems used in WLCDES.

Table 4.11 Software environment of WLCDES

Equipment	Languages	Applications
Micro VAX II	FORTRAN™	Analysis, optimization, evaluation
Micro VAX II	CommonLisp™	Conceptual design, evaluation
Micro VAX II	Pascal™	Modeling and simulations
GPX	UIS™	Geometric graphics

So far, two working versions of WLCDES have been developed. WLCDES1.0 can be run on terminals without computer graphics support. It performs conceptual design, analysis and evaluation (except for layout design and simulation) for wheel loaders. WLCDES2.0 requires a GPX workstation™. It can accomplish all design tasks including layout design and kinematic simulation.

(b) Features of WLCDES

WLCDES is a large knowledge integration environment. It provides the following features:

- It is an integrated distributed intelligent design system for wheel loaders. Its functions include conceptual design, performance analysis, kinematic simulation, scheme evaluation, and decision-making.

- Its modularity enables system configuration so flexible that the knowledge base is easily expanded and modified by the end users, rather than the original developers.

- It can accomplish both multi-objective design tasks and uncertain multi-criteria design tasks.

- The integrated distributed intelligent system concept and mate-system approach provide an open architecture to organize the system. Such a configuration allows a user to modify the knowledge base at any level of the system. In this knowledge-based system, the numerical computing process is coupled with symbolic reasoning and graphics simulation to enhance system capability.

- It has an interface with commercial databases.

(c) Illustration

The following demonstration simply illustrates the whole design process of a model ZL50 wheel loader.

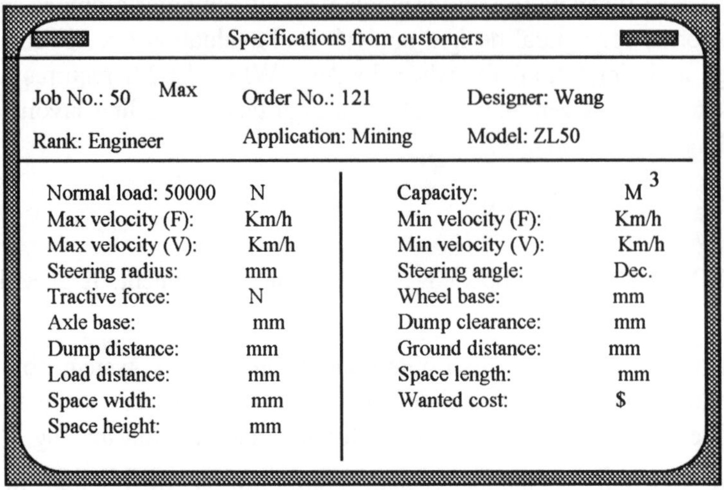

```
┌─────────────────────────────────────────────────────────────┐
│  ▓▓▓▓▓           Specifications from customers        ▓▓▓▓▓   │
│                                                               │
│  Job No.: 50  Max      Order No.: 121       Designer: Wang    │
│                                                               │
│  Rank: Engineer       Application: Mining    Model: ZL50      │
│─────────────────────────────────────────────────────────────│
│  Normal load: 50000   N    │  Capacity:              M ³      │
│  Max velocity (F):    Km/h │  Min velocity (F):      Km/h     │
│  Max velocity (V):    Km/h │  Min velocity (V):      Km/h     │
│  Steering radius:     mm   │  Steering angle:        Dec.     │
│  Tractive force:      N    │  Wheel base:            mm       │
│  Axle base:           mm   │  Dump clearance:        mm       │
│  Dump distance:       mm   │  Ground distance:       mm       │
│  Load distance:       mm   │  Space length:          mm       │
│  Space width:         mm   │  Wanted cost:           $        │
│  Space height:        mm   │                                  │
└─────────────────────────────────────────────────────────────┘
```

Figure 4.18 Task definition.

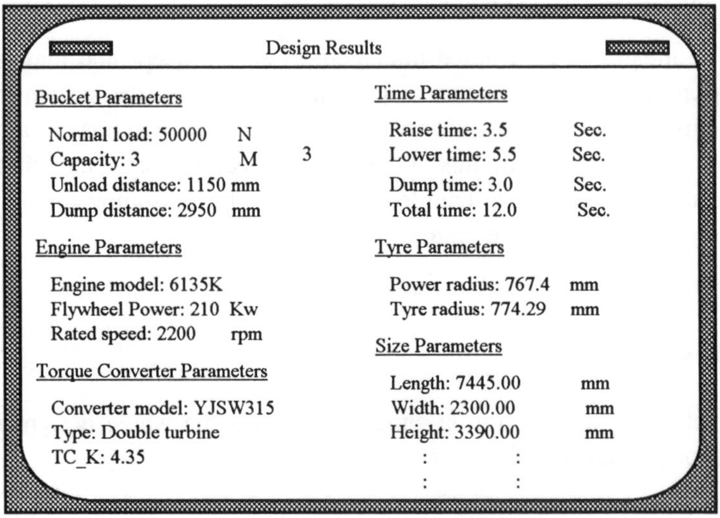

```
┌─────────────────────────────────────────────────────────────┐
│  ▓▓▓▓▓               Design Results                  ▓▓▓▓▓    │
│                                                               │
│  Bucket Parameters          Time Parameters                   │
│                                                               │
│  Normal load: 50000   N       Raise time: 3.5      Sec.       │
│  Capacity: 3          M ³      Lower time: 5.5      Sec.       │
│  Unload distance: 1150 mm     Dump time: 3.0       Sec.       │
│  Dump distance: 2950  mm      Total time: 12.0     Sec.       │
│                                                               │
│  Engine Parameters          Tyre Parameters                   │
│                                                               │
│  Engine model: 6135K          Power radius: 767.4  mm         │
│  Flywheel Power: 210 Kw       Tyre radius: 774.29  mm         │
│  Rated speed: 2200    rpm                                     │
│                             Size Parameters                   │
│  Torque Converter Parameters                                  │
│                               Length: 7445.00      mm         │
│  Converter model: YJSW315     Width: 2300.00       mm         │
│  Type: Double turbine         Height: 3390.00      mm         │
│  TC_K: 4.35                        :        :                 │
│                                    :        :                 │
└─────────────────────────────────────────────────────────────┘
```

Figure 4.19 Part design results of Model ZL50.

(1) First of all, a table of customer's requirements and specifications must be completed (Fig 4.18).

(2) After conceptual design, two acceptable schemes are generated. Each scheme has over 400 concept nodes that have their own attributes and parameters. Fig. 4.19 illustrates part parameters of scheme 1.

(3) In order to avoid collision and interference between the various parts of the working device, WLCDES carries out a layout design and motional simulation. The results are shown in Fig. 4.20 and Fig. 4.21.

Figure 4.20 Result of layout design.

Figure 4.21 Motional simulation of Model ZL50.

(4) Table 4.12 shows data and information from the analysis module. In WLCDES, the system analysis includes three parts:

- match analysis of engine, torque converter and gear box,
- stability analysis, and
- brake capability analysis.

The match analysis results of scheme 1 are shown in Table 4.12.

Table 4.12 Results of the match analysis

Number	First	level	Second	level	Third	level
	Force	Velocity	Force	Velocity	Force	Velocity
1	12839.8	0.000	4266.32	0.000	8412.63	0.000
2	10045.2	1.328	3267.80	3.585	6545.47	1.870
3	8916.72	2.652	2864.58	7.161	5791.49	3.736
4	7868.22	3.618	2489.95	9.769	5090.97	5.097
5	6571.59	4.646	2.26.65	12.544	4224.65	6,545
...	...	...	...	...	...	...

(5) According to the analysis results, an evaluation on the design quality is performed. Table 4.13 shows an estimation to the second level index of scheme 1.

Table 4.13 Evaluation to the second level index of scheme 1

Number	Indices	Score
1	Comprehensive performance	75.0
2	Reliability	64.0
3	Economy	71.0
4	Man–machine factor	60.0
5	Structure performance	66.8
6	Service	67.3
	General score	67.78

(6) The evaluation module finally assigns scheme 1 a comprehensive evaluation score 67.78. If the score for acceptance is 80.00, a redesign is required. Table 4.14 shows the optimization feedback information.

Table 4.14 Results of optimal feedback

Number	Indices	Previous values	Optimal values
1	Bucket volume (m^3)	3.00	3.00
2	Power rate (kw/Ton)	8.67	7.65
3	Drive force rate (N/cm)	543	578
4	Tractive force rate (N/cm)	709	695
5	Work time (s)	12.5	11.0
6	Maximum velocity (km/h)	34	31
7	Reliability of loaders (hours)	350	380
8	Technical utilization factor	86	90
9	Man-machine factor	84	90
10	Component reliability (hours)	9000	9300
11	Fuel consumption (gram/km h)	236	230
12	Maintenance cost ($)	1100	900
13	Manufacturing cost ($)	29k	25k
14	Hand operation force (N)	66	55
15	Foot operation force (N)	161.6	129.0

(7) According to the evaluation results, some parameters are changed in redesign. The partial results of redesign are shown in Table 4.15.

(8) The same design procedure is performed again to generate the second scheme. Table 4.16 compares two schemes on six aspects using the decision-making model (section 4.9). The index system selected has 33 index items on three levels. The evaluation result proves scheme 1 inferior to scheme 2, even though scheme 1 costs less than scheme 2. Moreover, evaluation at the highest level (general score) shows scheme 1 = 0.4386, while scheme 2 = 0.5614. The differences between the two general scores represents a distinction between the integrated

performance of the schemes. Thus, scheme 2 is selected as an optimal design.

Table 4.15 Partial results of redesign

Number	Indices	Previous results	Redesign results
...	... ...	...	...
15	Operation force 1 (N)	86.6	58.87
16	Operation force 2 (N)	147.0	148.85
17	Operability	66.4	76.16
18	Comfortability	51.4	76.83
19	Noise (dB)	98.0	87.03
20	Configuration	61.4	69.71
21	Safety protection	48.6	77.26
...	... ...	...	...

Table 4.16 Comparison of two schemes

	Functions	Reliability	Cost	Man–machine factor	Structure features	Services	General score
1	0.464	0.412	0.510	0.389	0.424	0.428	0.4386
2	0.536	0.588	0.490	0.611	0.576	0.572	0.5614

5

Integrated distributed intelligent simulation environment

5.1 Introduction

An integrated distributed intelligent simulation environment (IDISE) is developed for engineering layout design and animation. It uses three-dimensional color graphics running on a personal computer, performs geometric modeling, model-based reasoning, automatic generation of homogeneous transformation matrices, model assembling, path planning, and motional simulation. It can also be linked with other design software packages such as IDIDE (Chapter 4). IDISE can create various geometric models and parametric models, select an *effective model set* using its *model-based inference engine*, and evaluate all feasible layout schemes. The effective model set includes all acceptable layout design schemes, and is operated by the model-based inference engine based on the generated homogeneous transformation matrices and knowledge base. A *planning file* is generated automatically after a user provides positions and orientations (including six variables) of each body in a motional cycle. Finally, IDISE automatically creates slides of all bodies based on the planning file, and stores the slides in the hard disk as a *slide file*. At any time, the slide file can be executed for various motion displays on a computer screen. The above tasks or functions can be implemented interactively or automatically. The integrated distributed intelligent simulation environment does not require user programming and compiling for implementing a motional simulation of an engineering design.

On the other hand, IDISE offers an economic approach for industrial simulations such as testing design results, animating new robot designs, identifying collision between geometric bodies, planning robot motion paths, as well as serving as an education tool.

5.1.1 Robot simulation

Computer simulation techniques are frequently used in robot manufacturing applications to find collision and interference between the moving bodies, to plan optimal motion paths, and to test design results. It has also been used successfully in education to give students hand-on experience of programming a robot and designing a workcell. So far, several simulation packages for industrial robot programming and workcell design have been developed (Derby, 1982; Novak, 1984; Kovacs, 1985; Eydgahi and Sheehan 1991). A number of robot simulators have become available in the software marketplace (Yong *et al.*, 1985; Stauffer, 1984, Howie, 1983; Eversheim *et al.*, 1981), including GRASP™, McDonnell Douglas Robotic Software™, AutoSimulations™, AUTOPASS™ and ROBCAD™. In the area of robotics education, computer simulation software for PCs has also been developed and successfully used to teach fundamentals of robot motion and programming (White *et al.*, 1989; Denavit and Hartenberg, 1988).

These robot simulation packages above usually provide a set of models or a library of robots and objects (Eydgahi and Sheehan, 1991), as well as simulation tools that can be used to represent graphically a robot manipulator and its attendant equipment. So, they simulate, off-time, a robot performing some design tasks. The packages allow a design engineer to test several solutions for robotic cells before purchasing any equipment or investigating alternative uses of existing cells. Hence, these workplace design tools can help users to select and design robot systems, to reduce set-up costs and installation time, and to improve system performance. Some manufacturers claim that 'Engineers can design and lay out robot cells up to seventy percent faster through use of a robot simulator' (Stauffer, 1984).

5.1.2 Layout design simulation

Besides the applications of computer simulation in robot design and manufacturing systems, other fields of manufacturing such as mechanical product design, industrial process control, structure analysis and architecture design also frequently employ computer simulation to digitally implement an analogous working environment, events, object behavior in the real world, on a computer screen.

Computer simulation today is playing a more and more important role in engineering design to decrease design and manufacturing costs and improve design efficiency. The main purpose of simulation is to find the differences between a design model and a real-world object, and to reduce these differences through adjusting design variables.

In general, computer simulation in manufacturing industries can be divided into solid model simulation and process data simulation. The former includes geometric simulation (geometric modeling), body motional simulation (motion analysis) and body dynamic simulation (physical analysis). The latter involves production planning simulation (quality analysis) and process control simulation (system analysis) (Figure 5.1). Geometric modeling and motion simulation of robot and mechanical products using IDISE will be discussed in this chapter later.

Engineering design includes parameter design and structure design. The structure design also consists of two parts: structural configuration design and layout space design. IDISE can automate the space layout design of mechanical systems or products.

At present, most layout space designs in computer simulation are performed interactively. However, the interactive layout design method has the following obvious defects:

- Due to the limited capability and expert knowledge of a designer, only a small number of design layout schemes can be provided for a specific layout space issue. Then, an optimal or near optimal scheme is selected from these schemes through analysis and comparison. Thus, better schemes may be ignored.

• Layout space design is very complicated, and involves the representation of geometric models and space inference. Therefore, it is time-consuming and cost-ineffective to generate a practicable product model.

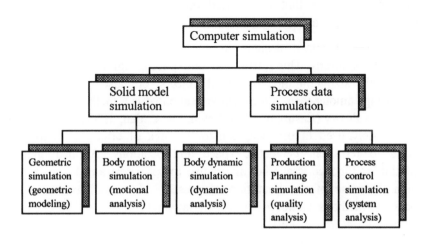

Figure 5.1 Computer simulation in manufacturing.

• Interactive layout design requires that a computer operator must possess much specialized domain knowledge and techniques. In fact, this kind of skilled personnel is lacking in many industrial sectors.

5.1.3 Automatic layout space design

Many research efforts have, for over 20 years, been devoted to automatic layout space design. These activities have focused on such major issues as layout design with or without restraints. A non-restraint layout space design does not require a user to provide geometric and topological relationships among the objects to be laid out. Generally, this is an

optimization problem (Markov, 1984), and can be solved by mathematical planning methods. However, problem solutions may be generated other than by using mathematical models when dealing with a complicated large-scale geometric planning problem because the geometric planning problem is usually time-consuming. To improve the computation efficiency, new algorithms that couple numerical computation with heuristic inference have been developed (Albano, 1980; Prasad, 1986). The key issue when using these methods is to establish heuristic evaluation functions that are carried out in accordance with specific conditions in application domains.

Most layout space designs in engineering are performed under constraint conditions. This restraint layout space design needs to consider not only the size limitation of layout objects, but also the topological relationships among them. One technique adopted is a methodology based on graphics theory (Eastman, 1970; Levin, 1964; Yan, 1988). However, it is not widely used in practice, due to the lack of effective computing models to represent geometric space and positions. Applications of artificial intelligence is another alternative to solve restraint layout space design. Because of the limitation of present AI technology, symbolic processing methods cannot generate very satisfactory results (Wang *et al.*, 1990b; Liu *et al.*, 1989). The objective of developing IDISE is to introduce integrated distributed AI technology into interactive layout space design systems in order to strengthen the functions of existing CAD systems and to enhance the efficiency and quality of layout space designs, rather than to replace the interactive design functions.

5.2 IDISE configuration

The integrated distributed intelligent simulation environment (IDISE) is so developed that an intelligent module is added into the interactive layout design system (LDS) (Li *et al.*, 1990), as described in Figure 5.2. LDS (Layout design system) is an interactive 3D graphics system, which can be used to create geometric solid models, interactively assemble the models, automatically check out collision (interference) between geometric bodies, and simulate the motion of mechanisms. This system has been employed in mechanical designs for robots, wheel loaders, machining centers and fixture designs. However, some shortcomings have been found when LDS is applied:

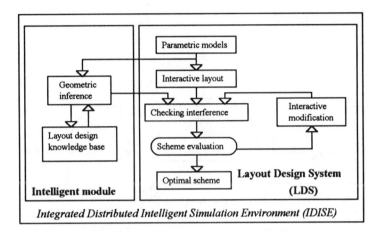

Figure 5.2 Software structure of IDISE.

- Operation of LDS requires users to be well trained in both application domain and computer graphics.

- Users sometimes may not fully consider all technical requirements and restraints, so it is difficult to obtain all feasible schemes. The restraint conditions include not only geometric constraints, but also functional and topological constraints.

- The efficiency of interactive layout space design is rather low.

To enhance the capacity and efficiency of LDS, an integrated distributed intelligent simulation environment (IDISE) based on LDS has been developed. IDISE partially accomplishes automation of layout space design and mechanism motional simulation. IDIDE acts in the following manner:

(1) to interactively generate the parametric 3D models and store the basic models in the model base (see section 4.4 and Figure 4.4);

(2) to choose a set of effective models from the model base using model-based reasoning, which can satisfy the specific requirements and application environment;

(3) to resolve the effective models into several practical layout design schemes;

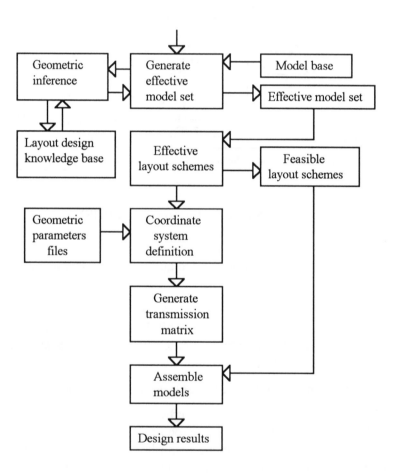

Figure 5.3 Procedure flow diagram of intelligent layout design module.

(4) to automatically conduct homogeneous transformation matrices among geometric models based on position and orientation parameters in space;

(5) to manipulate these matrices and the basic parametric models to generate practical layout drawings; and

(6) to generate animation slides and simulate mechanism motion of each layout scheme to check whether geometric restraints are satisfied. If a layout scheme does not meet the requirements, it is modified through an interactive 3D graphics system. If all layout schemes meet the requirements, an analysis and evaluation of each scheme is carried out to select an optimal or near optimal scheme. Figure 5.3 demonstrates the procedure flow diagram of the intelligent layout design module.

5.3 Model definition and data structure

In IDISE, the technique of parametric modeling is adopted. Each model has several variables or parameters that determine the position, size, and orientation of a body. When these geometric parameters are changed, the size, position and orientation of the model are also changed, but the shape of the model remains the same.

In conventional geometric modeling, every model contains only its geometric information. The attribute values and topological information for the models are very important in automatic layout space design, but, in conventional modeling, are neglected. An IDISE's model includes the following information:

- geometric information that determines the shape, size and position of a model;

- attribute values that include the name, order, function, color, line style of a model and so on;

- topological information that represents the coordinate and assembly relationships among models, as well as the model with space invariance.

Figure 5.4 describes the definition and data structure of parametric models.

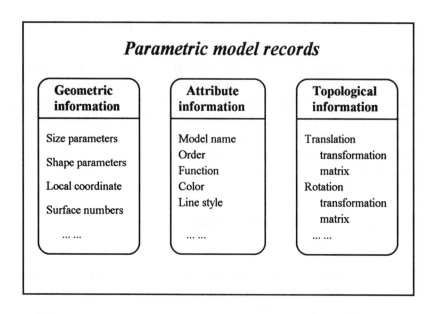

Figure 5.4 Definition and data structure of parametric models.

5.4 Model base

A mechanical system (a set of parts or components) may consist of many subsystems, which are further divided into their respective subordinate subsystems. Thus, the system can be continuously subdivided into basic models (elements). The models are not arbitrarily heaped up, but are organized into a hierarchical structure as shown in Figure 5.5. Assume M_i is a subset of model M, then

$$M_i = \{m_{in} \mid A_{ik}(M_{ik}) \quad n = 1, 2, .., N; \ k = 1, 2, .., K\} \tag{5.1}$$

where m_{in} is the nth model of model subset M_i, M_{ik} is a subset of M_i (i.e. $M_{ik} \in M_i$), and A_{ik} expresses the relationship of position and orientation between models; that is, it is a homogeneous transmission matrix. Similarly, a mechanical system can be described as follows:

$$M = \{m_n | A_k(M_i) \quad i = 1, 2, .., I; k = 1, 2, .., K\} \qquad (5.2)$$

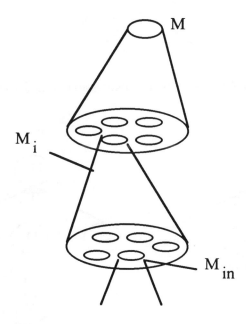

Figure 5.5 Hierarchical structure of a mechanical system.

Since a model set is an ordered, hierarchical structure, it can be expressed by an 'AND/OR' tree as shown in Figure 5.6. The 'AND' node shows that all models connected to the node in a model set must be included to assemble the model set, while the 'OR' node only uses one model to realize the same objective.

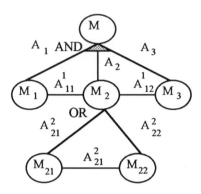

Figure 5.6 OR/AND tree expression of model set.

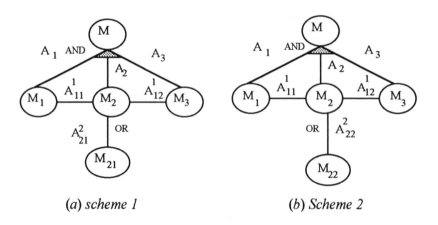

(a) scheme 1   (b) Scheme 2

Figure 5.7 Examples of effective model set conversion.

This model set (or model base) contains all the possible layout schemes. That is to say, at every 'OR' node, several layout space schemes can be resolved. For instance, the model set in Figure 5.6 can be resolved into schemes (a) and (b) of Figure 5.7.

Two problems are encountered above, the first one is how to prune a model set (tree) in terms of the requirements and geometric restraints so that the set can be transferred into an effective model set. Another is how to resolve an effective model set into several feasible layout schemes.

5.5 Model-based inference

5.5.1 Knowledge representation

An effective model set can be performed through model-based inference. First, the specifications and geometric restraints are linked to every node of a model set (i.e. an 'OR/AND' tree). Those specifications and restraints are premise conditions to trigger each node. After doing this, a model-based inference network for layout space design is developed, as shown in Figure 5.8.

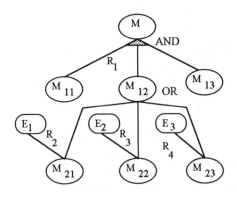

Figure 5.8 Model-based reasoning network.

In Figure 5.8, E_i represents a restraint condition. If conditions E_1 and E_2 are true, and E_3 is false, M_{23} can be pruned away. This task is carried out by the model-based inference engine. As an example, the model set described in Figure 5.8 is turned into the set expressed in Figure 5.6. The transformed set is an effective model set.

The model-based inference network is also the knowledge base of the intelligent layout module. In order for the logic relationship among models in Figure 5.8 to remain the same as in Figure 5.6 after reasoning, a special form of knowledge representation and organization is used (see section 4.5.2). The premise conditions of rules in model-based inference contain many parameter variables, matrices and numerical computation procedures. The conclusions of rules are usually the functions or operations, rather than declarations and interpretations. For example, the knowledge base of model-based inference consists of three parts:

Rules: The rules for Figure 5.8 can be expressed as follows:

R_1: (IF M THEN (M_{11} M_{12} M_{13}))

R_2: (IF (AND E_1 M_{12}) THEN M_{21})

R_3: (IF (AND E_2 M_{12}) THEN M_{22})

R_4: (IF (AND E_3 M_{12}) THEN M_{23})

Relationship between rules and nodes: The relationship between rules and nodes can be divided as *forward rules* and *backward rules* (see section 4.5.2). Here the relationships are also represented by *Before* and *After* lists. Referring to Figure 5.8, the two lists may be expressed as:

Before: ((M_{11} (R_1)) (M_{12} (R_1)) (M_{13} (R_1))

((M_{21} (R_1 R_2)) (M_{22} (R_3 R_1)) (M_{23} (R_4 R_1)))

After: ((M (R_1 R_2 R_3 R_4)) (M_{12} (R_2 R_3 R_4)))

Record of goal nodes: A *goal node list* records the nodes that have no any subnodes. A goal node in the goal list expresses a basic geometry model or element that cannot be separated. For instance, a goal node is the *leaf node* in Figure 5.8, and its goal nodes are:

Goal: (M_{11} M_{13} M_{21} M_{22} M_{23})

The above explanation can be described by

$$MT = (M, R, Q) \tag{5.3}$$

where

MT = stands for a model-based reasoning network,

$M = \{C_i\}$ (i=1, 2, ..., I), and is a set of model nodes,

$R = \{r_j\}$ (i=1, 2, ..., J), and is a set of rules, and

Q = {After, Before}, and represents the relationship between nodes and rules (forward rules or backward rules).

5.5.2 Model-based inference algorithm

To introduce the model-based inference algorithm, let us define:

Stack = to store expanded nodes,

Facts = to store restraint conditions,

Middle = to store negative premise conditions, and

Final = records results (effective model set).

The model-based inference algorithm consists of the following steps:

(1) Given M_i (a system/product to be designed).

(2) Put M_i in Stack and Final.

(3) Check whether Stack is empty. If yes, print the effective model set stored in Final, and exit.

(4) Take M_i out of Stack. Check all its backward rules in After list, and backward nodes associated with the rules (i.e. all subnodes and premises of M_i).

(5) If M_i has a subnode (i.e. M_i is not in the goal list),

(i) If this subnode is not a goal node, push it into Stack, then go to (3).

(ii) If this subnode is a goal node, put it in Stack, continue.

(6) If M_i has no subnodes or M_i is a goal node, verify the premises of forward rules associated with M_i:

(i) Try to match the premises with restraints in Facts. If M_i is confirmed, put it in Final, then store the triggered rule and go to (3). If not, go to (ii) (6).

(ii) Consult a user, and put the information provided by the user into Facts (positive facts) and Middle (negative facts). If M_i is confirmed, put M_i into Final, and then, store the triggered rule and return to (3),

(iii) If M_i is not confirmed, put the negative conclusion in Middle and give up M_i, then, return to (3).

5.6 Separation operation of effective model set

An effective model set contains all the feasible layout schemes. The separation operation can divide the effective model set into several independent layout schemes (subsets). The separation operation is often carried out at 'OR' nodes. Suppose that an effective model set (or an *effective tree*) has n 'OR' nodes on a tree and t branches at a node, then the number of layout schemes is given as

$$NM = \prod_{i=1}^{n} t_i \qquad (5.4)$$

For example, Figure 5.6 displays an effective model set, so the model-based inference result stored in Final is

Final $= (M (M_{11} (M_{12} (M_{21} M_{22})) M_{13}))$.

This can be resolved into two layout schemes:

Scheme 1 $= (M (M_{11} (M_{12} (M_{21})) M_{13}))$, and

Scheme 2 $= (M (M_{11} (M_{12} (M_{22})) M_{13}))$.

Before introducing the separation operation algorithm of effective model set, an effective model set is defined as

$$EMS = (EM, ER, Q). \qquad (5.5)$$

where

$EM = \{em_i\}$ (i=1, 2, ..., I), it is an effective model set,

$ER = \{er_j\}$ (j=1, 2, ..., J), it is a triggered rule set, and

$Q = \{After, Before\}$, it represents the relationship between an effective node and triggered rules.

The separation operation algorithm (*LS* stores the feasible layout schemes) is described as follows:

(1) Put the effective model set in Final into Stack.

(2) Verify whether Stack is empty. If yes, display data in *LS* and exit. Otherwise,

(3) Take L_i out of Stack, and search 'OR' nodes from the bottom of L_i to the top,

   (i) If an 'OR' node is found, and has more than two branches, L_i should be divided into several sublists (subsets) at this node. Put the sublists in Stack, and go to (2).

   (ii) If no 'OR' node is found, put L_i into *LS* and then go to (2).

5.7 Generating a homogeneous transformation matrix

For most mechanical products or systems, a model set has a hierarchical structure. Therefore, the homogeneous transformation matrices that represent the relative positions and orientations between models also have a hierarchical configuration. Each level can be viewed as a *closed loop*. Models inside the closed loop and the coordinates between closed loops can be transferred into each other. In IDISE, each model or subsystem has its own coordinates and can be assembled at a specified position through the homogeneous transformation which describes the relative position and orientation between these coordinates. Here, the homogeneous transformation matrices that express the assembly relationships are shown as matrix $\tilde{A}$ in Figure 5.9. These $\tilde{A}$ matrices have hierarchical structures. Each layer is a closed loop (Figure 5.10). The models in the closed loop and the coordinate systems between the closed loops can be transferred. Thus, each model can be fixed in its coordinate system, and the $\tilde{A}$ matrices describe the relationship between coordinate systems.

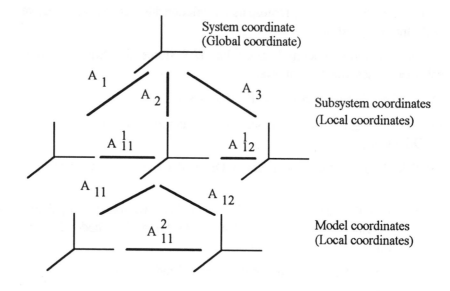

Figure 5.9 Homogeneous transformation between coordinate systems.

Assume $\tilde{A}_1$ describes the position and orientation of the first geometric model, while $\tilde{A}_2$ represents the position and orientation of the second geometric model related to the model. The position and orientation of the second model related to the *global* coordinate system can be expressed:

$$\tilde{T}_1 = \tilde{A}_1 \tilde{A}_2 \tag{5.6}$$

Similarly, the position and orientation of the nth model ($\tilde{T}_n$) related to the global coordinate system can be expressed as:

$$\tilde{T}_n = \prod_{i=1}^{n} \tilde{A}_i \tag{5.7}$$

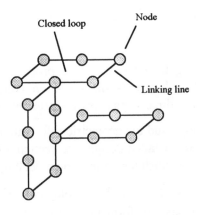

Figure 5.10 Assembly relationship among geometric models.

In the equation above, $\tilde{A}_i$ stands for the position and orientation of the geometric model i related to that of the model i-1. Usually, such single closed loop structures do not exist in a product or system, because most of them often consist of subsystems or branches at every node. The nodes in branches are also connected with the nodes in the main closed loop as shown in Figure 5.10. Here, the linking lines (arcs) represent the coordinate relationships of geometric models or subsystems, while the nodes express homogeneous transformation among subsystems or models.

This structure is complex. Actually, it is a recursive structure, and can be automatically generated by computers. The inputs from a user are only the relative position and orientation matrix (i.e. matrix $\tilde{A}$) of a geometric model, as well as the coordinate features between geometric models.

In IDISE, the coordinate relationships (or features) are described in terms of *Part of, Attachment, Constrain* or *Assembly* (Li *et al.*, 1990). It is very easy for engineers to provide the coordinate relationships between geometric models because the relationships are directly determined by the special domain knowledge and application environment.

5.8 Model assembly

Essentially, engineering layout design is a process of model assembly. Each of the feasible schemes is displayed on computer screen. This process is carried out in three steps using a Cartesian coordinate system:

(1) Geometric modeling: Add the geometric information obtained from the parameter design (or detailed design, see section 4.6) to each effective model set, then accomplish the parametric modeling of all bodies.

(2) Generating matrices $\tilde{A}$: Add the topological information into each effective homogeneous transformation matrix $\tilde{A}_i$.

(3) Model assembly: Put each effective model on its relevant position and orientation through matrix operation. For instance, $\tilde{A}_m$ is a translation matrix; T_x, T_y and T_z are translation parameters in the direction of X, Y, Z coordinates respectively; $\tilde{A}_{vx}$ $\tilde{A}_{vy}$ $\tilde{A}_{vz}$ are the corresponding rotation matrices round X, Y and Z coordinates; and θ stands for a rotation angle. These matrices are presented below:

$$\tilde{A}_m = \begin{vmatrix} 1 & 0 & 0 & 0 \\ 0 & 1 & 0 & 0 \\ 0 & 0 & 1 & 0 \\ T_x & T_y & T_z & 0 \end{vmatrix} \qquad \tilde{A}_{vx} = \begin{vmatrix} 1 & 0 & 0 & 0 \\ 0 & \cos\theta & \sin\theta & 0 \\ 0 & -\sin\theta & \cos\theta & 0 \\ 0 & 0 & 0 & 1 \end{vmatrix}$$

$$\tilde{A}_{vy} = \begin{vmatrix} \cos\theta & 0 & -\sin\theta & 0 \\ 0 & 1 & 0 & 0 \\ \sin\theta & 0 & \cos\theta & 0 \\ 0 & 0 & 0 & 1 \end{vmatrix} \qquad \tilde{A}_{vz} = \begin{vmatrix} \cos\theta & \sin\theta & 0 & 0 \\ \sin\theta & \cos\theta & 0 & 0 \\ 0 & 0 & 1 & 0 \\ 0 & 0 & 0 & 1 \end{vmatrix}$$

For the example in Figure 5.7 (a), the assembly operation on subsystem M_2 is carried out as follows:

$$M_2 = M_{21}\tilde{A}_{m21}\tilde{A}_{v21} \tag{5.8}$$

Similarly, we can obtain M_1 and M_3, thus, the final layout result of system M can be calculated:

$$M = M_1\tilde{A}_{m1}\tilde{A}_{v1} + M_2\tilde{A}_{m2}\tilde{A}_{v2} + M_3\tilde{A}_{m3}\tilde{A}_{v3} \tag{5.9}$$

Here, M is one of effective layout schemes. $\tilde{A}_{m1}$, $\tilde{A}_{m2}$, $\tilde{A}_{m3}$, $\tilde{A}_{v1}$, $\tilde{A}_{v2}$, and $\tilde{A}_{v3}$ are determined by the geometric and topological information of assembly. Suppose N is a scene transformation matrix and W is a viewpoint transformation matrix, then

$$M_2 = M_{21}\tilde{A}_{m21}\tilde{A}_{v21}NW \tag{5.10}$$

5.9 Fast removing of hidden lines and surfaces

In order to fast remove hidden lines and surfaces, over two hundred algorithms have been proposed. These algorithms can be essentially classified into two categories. One is the object-space algorithm, and the other is the image-space algorithm (Angell and Griffith, 1990). The efficiency of these algorithms is related to a specific problem (Harrington, 1987). To enhance the efficiency of these algorithms, two methods are usually adopted: (i) to use the properties of succession and local relevance possessed by a body and its image, and adopt an incremental algorithm to quicken the process. (ii) To reduce the number of explorations and comparisons to the lowest possible limit.

Recently, the research on fast removing hidden lines and surfaces has focused on two aspects: one is the study of body relevance, while the other is the study of the *priority degree* of a body in graphical display. The study of body relevance affects the algorithm quality of priority degree. The present algorithms for body relevance mainly focus on geometric relations between bodies and overlook the physical relationships among models.

Painter's algorithm (or the back-to-front method) (Angell and Griffith, 1988) is widely used in commercial graphics software development. During the process of painting in oils, a painter always starts from the background, and then sequentially paints new graphs along the background. When an oil graph has to be overlapped, a new graph automatically covers the old one. A graph display is determined by painting sequence. This is Painter's algorithm, i.e. the 'back-to-front' rule. In IDISE, the basic idea of Painter's algorithm is used to quickly remove hidden lines and surfaces of a polyhedron. According to the *depth values* of polygons (e.g. the distances between from a viewpoint to bodies), the polygon's priority degrees are decided. Thus their *display order* is arranged. In terms of the 'back-to-front' rule, polygons with lower priority degree are first displayed. Because the priority degree of the former treated polygons is always lower than that of the latter treated ones, polygons with a higher priority degree will automatically cover polygons with a lower priority degree when overlapping occurs.

Generally, a polyhedron is a closure formed by many surfaces. Each surface is a polygon. There are two kind of surfaces, denoted as S_a and S_b. When polygons are enlarged or extended, if the surfaces of a polyhedron can be separated, the surfaces are S_a surfaces. Otherwise, they are S_b surfaces. Obviously, only non-convex polyhedrons contain S_a surfaces, while convex polyhedrons do not contain S_a surfaces. Three important properties are summarized below:

Property 1: S_a surfaces cannot hide S_b surfaces. The reverse is not true.

Property 2: There exists no hidden relationship between S_b surfaces.

Property 3: There exists no hidden relationship between the uncovered parts of S_a surfaces (related to S_b surfaces).

Using the above three properties, we can remove the back surfaces of a polyhedron, and then clip that part of S_a surfaces hidden by S_b surfaces. Since there is only a property level between S_a and S_b surfaces and no hidden relationship between polygons in S_a and S_b, the priority degree is zero. The priority degrees in all the S_b surfaces are higher than those in S_a surfaces. If S_a surfaces are displayed first, and S_b surfaces afterwards, the hidden lines and surfaces between the surfaces of a single body will be removed.

Now, let us discuss the relationships between geometric bodies in terms of hidden lines and surfaces. When a mechanical product or system is assembled, some fixed relationships between components or parts are established. Using these relationships, we can determine the relative position and orientation of a body in space, or its position and orientation in the global coordinate system. Thus, the order relationship formed by the depth values can be obtained. Meanwhile, the priority degree can be decided easily based on the depth values.

If a polygon in a body is inserted by another body, then there exists an *intercross relation* between the bodies. Otherwise, there exists a *connection relation* between them. For bodies with the connection relation, the hidden relationship between a *front-body* and a *back-body* can be determined by their depth values. Therefore, the removal of hidden lines and surfaces becomes very simple. What we need to do is to display a back-body, then display a front-body. For the bodies with intercross relation, the removal of hidden line is rather complex, and set operation algorithms have to be used.

5.10 Collision and interference testing

Testing collision and interference between geometric models is often used in layout space design to check whether a body is intersected by another. Obviously, the intersection can be identified by set operations. However, for a product with many geometric bodies, it is very difficult to find intersections directly. Moreover, it is impossible to find the trend in the occurrence of collisions in the real-time environment. In IDISE, collision-testing is carried out in three steps to enhance the efficiency:

- using the assembly relationships of geometric bodies to identify bodies which may result in collision,

- using the J function (Li *et al.*, 1990) to find the shortest distance between bodies and to identify the bodies which produce collision, and

- Finding intersection locally and examining the degree of interference.

5.11 Geometric modeling

Geometric modeling mainly addresses how to represent and generate 3D geometric information of bodies. In IDISE, two technical issues are handled: (i) shape definition: provides users with a tool to accurately describe the information of geometric shapes; (ii) data structure: enables data structure transformation and data exchange easily.

IDISE provides three types of solid-modeling methods and a user-friendly interface. Hence, users can interactively perform solid-modeling and parametric modeling. The following methods have been developed for geometric modeling:

5.11.1 2–1/2D geometric modeling

During 2–1/2D graphic operation, IDISE requires a user first to construct a 2D graph, and then to provide the data about the sweeping length, or the position of the rotating axis, as well as other relevant information. Finally, the system can build a 2–1/2D solid model by numerical computation. These generated bodies can serve as basic elements for set operation. In IDISE, this process includes two stages: (i) producing a simple, transitional data structure, and (ii) transforming this data structure into the final data structure. The two-stage generation process makes the transitional data structure similar to all basic elements, thus it is convenient for graphics database management.

5.11.2 Set operation

IDISE uses the boundary-representation method (Angell and Griffith, 1990) to describe geometric bodies and develop such set operations as *union*, *intersection* and *difference* operations based on the representation. Two bodies to be operated on may be polyhedrons. Generally speaking, a body can be represented by a point set in Euler space, but it is not true that any point set can represent a body. In mathematics, a body is defined as a closed regular set in 3D Euler space. Since a body can be defined by a point set, the set operations between two bodies can be defined by the set operations of the point sets (Harrington, 1987).

5.11.3 Local operation

A local *reforming operation* is widely used to carry out processing locally and to generate the shapes required by the users. In IDISE, users can make local modifications to any surface, prismatic edge and vertex of basic elements, as well as construct solid models with similar shapes to real-world bodies. Through a polishing process, IDISE automatically produces the control vertex of freedom surfaces and connects them into required shapes.

5.12 IDISE functions

This environment enables implementation of geometric modeling integration, model-based inference, assembly, motional planning, and animation, as well as the utilization of geometric and space knowledge for automatically performing layout space design and simulation. IDISE provides the seven main modules as shown in Figure 5.11, and described below:

[File&Scrn] This can be used to manage geometric data files, to change windows, as well as to modify scales, viewpoint and light source.

[Modeling] Its functions are to create all solid geometric models and parametric models, modify and display the models, as well as manage and store the models.

[Reasoning] Through model-based inference, the module can choose effective models from basic parametric models, set up an effective model set, divide the set into several feasible layout design schemes (or subsets), as well as generate homogeneous transformation matrices based on parameter design and parametric modeling.

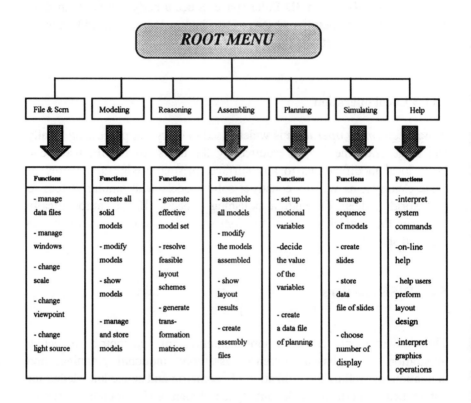

Figure 5.11 Functions of IDISE.

[Assembling] This is employed to assemble all effective models, modify the assembled models, display the layout results, and create assembling files.

[Planning] A designer can use the module to set up motional variables interactively, select the values of variables, and create a data file of motional path planning.

[Simulating] This can decide the sequence of all models to be displayed (e.g. compute depth values or priority degree), generate all slides, store a data file of slides, choose cycle number, implement the mechanism animation, and check collision and interference of bodies.

[Help] The module displays an on-line help window to explain all IDISE commands.

IDISE integrates various functions and software structures to implement engineering layout space design and simulation. Thus, it has the following features:

- It provides several methods to construct 3D solid models and enables these models to be modelized and parametrilized.

- It can automatically generate several effective layout schemes through the model-based inference engine.

- It can interactively assemble models or elements, then automatically set up data structures.

- It can automatically generate homogeneous transformation matrices and process 3D computer graphics.

- It implements tests for collision and interference between bodies.

- It can simulate and evaluate the designed schemes.

5.13 Examples

The main menu of IDISE (root menu) is shown at the right of the screen. When a choice is made from the main menu, a window containing all sub-menu commands appears on the right-hand side of the screen. The remaining area of the screen is the workspace where all solid models, assembled models, slides and motion models appear. Menu selection can be made easily with a mouse or with a keyboard.

IDISE is written in Pascal™ . It runs on PC-486 or compatible machines with 640K RAM, DOS5.0™ or higher, an EGA™ or VGA™ graphics monitor and a mouse. Because of the large amount of floating point calculations required to perform set operations and other transformations, a floating point co-processor with a speed of 33 MHz or higher is required. IDISE takes advantage of the EGA graphic capability using a 640×350 resolution with 16 colors to create realistic three-dimensional solid models. Real-time hidden page animation is performed to show model motion. IDISE uses a mouse as its primary input device. The keyboard is only required for a few special tasks. The mouse makes menu selection and data input much faster and more simple.

IDISE can be used for solid-modeling, assembly, path-planning and simulation in industrial manufacturing. The following industrial applications have been developed using IDISE:

(1) Conceptual layout design: IDISE, as an independent system, has been integrated with IDIDE (integrated distributed intelligent design environment) to accomplish scheme layout design tasks in the conceptual design of mechanical products. Examples can be found in Figure 4.14 through Figure 4.21 in Chapter 4.

(2) Combination fixture selection and assembly: IDISE obtains information from the expert system of combination fixture selection to implement the assembly automation of models and test design results. Figure 5.12 through Figure 5.14 demonstrate part of the assembly results based on geometric information from the expert system of combination fixture selection.

(3) Robot simulation: IDISE can be used in robot design and engineering education to implement robotic design, simulation and education. Figure 5.15 (a) and Figure 5.15(b) show examples.

(4) Machining center simulation: IDISE can be also employed in a CIM environment to realize animation of a machining process. The example shown in Figure 5.16 is a very complex system with 156 basic models.

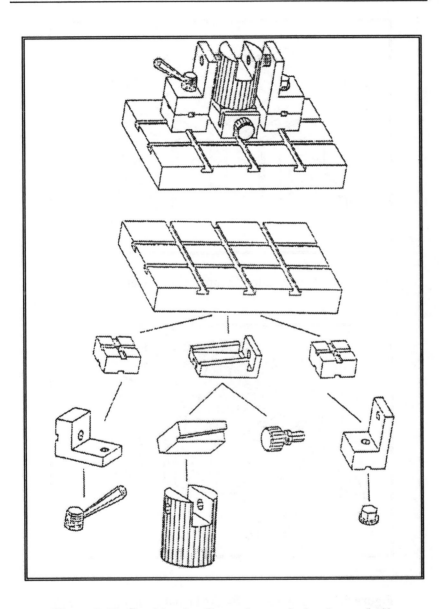

Figure 5.12 Combination fixture layout design (example 1).

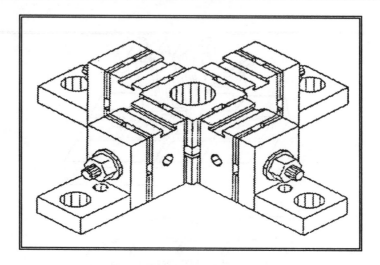

Figure 5.13 Combination fixture layout design (example 2).

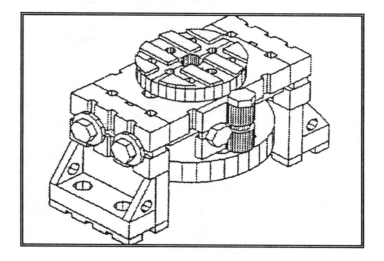

Figure 5.14 Combination fixture layout design (example 3).

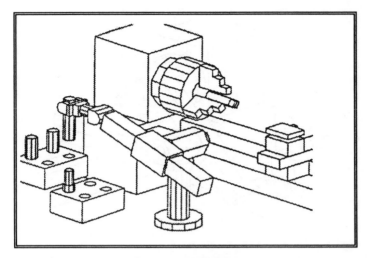

Grasp a workpiece

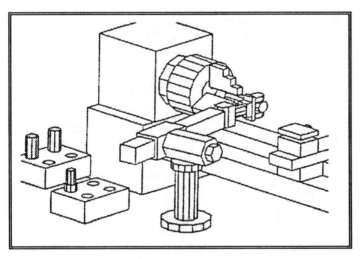

Install the workpiece

Figure 5.15(a) Robot assembly design and simulation.

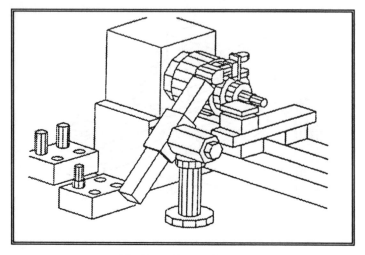

Machine the workpiece

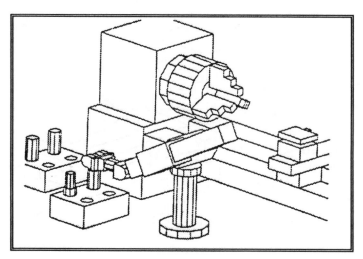

Loosen the workpiece

Figure 5.15(b) Robot assembly design and simulation.

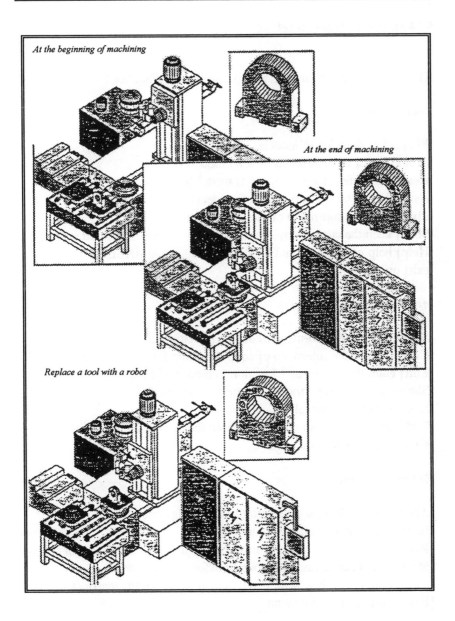

Figure 5.16 Layout design and simulation of machining center.

5.14 IDISE instruction set

IDISE menu system presents the following main functions:

Root menu:

[File] select data file of models
[Model] enter the sub-menu of solid modeling
[Reasoning] enter the sub-menu of model-based inference
[Assemble] enter the sub-menu of models assembling
[Plan] enter the sub-menu of motion planning
[Shell] keep the current environment and enter DOS operation system.
[Help] help users to operate his/her own models in IDISE.
[exit] quit from IDISE

Modeling menu:

[Screen] enter the sub-menu of screen management
[Create] enter the sub-menu of solid modeling for new geometric models
[Edit] enter the sub-menu of edition to modify old geometric models
[Show] show the solid models created or modified
[Delete] remove a model
[Fill] add a model from another file into the current file of models
[Exit] return to the root menu

Reasoning menu:

[Acquisition] add new rules into the knowledge base
[Inference] generate effective models
[Resolve] resolve the effective models into several feasible layout schemes
[Matrices] generate homogeneous transformation matrices
[Auto] automatically assemble the models based on the matrices
[Exit] return to the root menu

Assembling menu:

[File] select a data file of models assembled
[Screen] enter the sub-menu of screen management
[LoadFit] load an assembly data relative to the models assembled
[Fit] enter the sub-menu of assembling models
[Block] enter the sub-menu of block operation
[Exit] return to the root menu

Planning menu:

[File] select the data file of motion planning
[SetVar] enter the sub-menu of setting up motion variables
[Animate] enter the sub-menu of motion planning
[Sldshow] show the slides stored as the file of slides
[Exit] return to the root menu

Creating model menu:

[<- ->] adjust the position of the window
[Scale] change the size of model
[Origin] modify the original point of Cartesian coordinates
[Spline] enter the sub-menu to create spline curves
[Arc] enter the sub-menu to create arcs
[Skip] pick up "the pen" and move to the next vertex
[Undo] abandon the last vertex
[Edit] enter the editing sub-menu to modify vertex position
[Draw] show 3D solid model after sweeping or rotating 2D model
[Name] name the model
[Color] choose the color of model
[Options] select parameters such as sweeping thickness
[Save] save the current model in the hard disk
[Load] load the old model data files
[New] clear all old 2D vertices to create new model
[End] return to the modeling menu

Creating spline menu:

[Undo] abandon the last vertex
[B-spline] generate B-spline not to pass all vertices
[CB-spline] generate B-spline to pass all vertices
[Lagrange] create Lagrange curve through all vertices
[Bezier] create Bezier curve
[Abort] return to the creating model menu

Editing menu:

[Previous] move the previous vertex to be modified
[Next] move the next vertex to be modified
[EditPnt] modify the present vertex
[Insert] insert new vertex before the present one
[Delete] remove the present vertex
[Radius] change an angle to an arc
[End] return to the modeling menu

Fitting menu:

[ScrnMana] enter the sub-menu of screen management
[Fitted] show the relationship tree of models assembled
[ReadMDL] read a model
[Slt Base] select the assembling coordinate (global or local)
[Change U] enter the sub-menu to modify the U (motion) coordinate
[Change V] enter the sub-menu to modify the V (physical) coordinate
[Fit] assemble a model according to its position and orientation
[Del Back] delete the last assembled model
[Edit] re-assemble the special model
[Redraw] re-show the assembled model
[Return] return to the modeling menu

Block operation menu:

[ScrnMana] enter the sub-menu of screen management

[Fitted] show the relationship tree of models assembled
[Slt Base] select the assembling coordinate
[Change U] enter the sub-menu to modify the U coordinate
[Change V] enter the sub-menu to modify the V coordinate
[SaveTree] save the relationship tree or the relationship sub-tree
[Array] copy a component of array or rotation
[Copy File] read a data file from the disk
[Copy Mdl] copy a body or component
[Delete] delete the component index (can be resumed)
[Recall] resume all indices deleted
[Pack Del] remove all indices (can not be resumed)
[Redraw] re-show the assembled model

Copy model:

[Change U] enter the sub-menu to modify the U coordinate
[Change V] enter the sub-menu to modify the V coordinate
[Verify] confirm the model to be copied in the present coordinate
[Copy] copy a model or component
[Abort] return to the block operation menu

Change U (motion) coordinate:

[Fix V] determine if fix the V coordinate
[1-Move X] move the U coordinate along the base coordinate's X
[2-Move Y] move the U coordinate along the base coordinate's Y
[3-Move Z] move the U coordinate along the base coordinate's Z
[4-Turn X] rotate the U coordinate round-right the base coordinate X
[5-Turn Y] rotate the U coordinate round-right the base coordinate Y
[6-Turn Z] rotate the U coordinate round-right the base coordinate Z
[7-Turn1 X] rotate the U coordinate round X (change the U position)
[8-Turn1 Y] rotate the U coordinate round Y (change the U position)
[9-Turn1 Z] rotate the U coordinate round Z (change the U position)
[==>BS] set up the U coordinate equal to the base coordinate
[==>V] set up the U coordinate equal to V coordinate
[Return] return to the block operation menu

Change V (physical) coordinate:

[1-move] move the V coordinate along U coordinate's X
[2-move] move the V coordinate along U coordinate's Y
[3-move] move the V coordinate along U coordinate's Z
[4-Turn X] rotate the V coordinate round-right U coordinate X
[5-Turn Y] rotate the V coordinate round-right U coordinate Y
[6-Turn Z] rotate the V coordinate round-right U coordinate Z
[7-Turn1 X] rotate the V coordinate round X (change the V position)
[8-Turn1 Y] rotate the V coordinate round Y (change the V position)
[9-Turn1 Z] rotate the V coordinate round Z (change the V position)
[==>U] set up the V coordinate equal to U coordinate
[Return] return to the block operation menu

Setting up motion variables:

[ScrnMana] enter the sub-menu of screen management
[LoadFit] loading the data file of an assembled model
[File] open a file to store the data of motion variables
[InputVar] input a set of the motion variables
[Draw] show the assembled model
[Draw Nxt] input next data set and show the assembled model
[Batch] read all data and show all positions of the assembled model
[ViewCut] adjust the position of the window
[DrawSLDs] read all data from the file and show all slides
[ShowSLDs] continuously show all slides stored in the data file
[Return] quit and return to the planning menu

Editing and modifying motion variables:

[Select] choose a body relative to the present variable
[0-Param] set up a variable of the body
[1-Mov X] change the variable along the U's X
[2-Mov Y] change the variable along the U's Y
[3-Mov Z] change the variable along the U's Z
[4-Rot X] change the variable round-right the U's X

[5-Rot Y] change the variable round-right the U's Y
[6-Rot Z] change the variable round-right the U's Z
[CheckVar] check the changed variable
[Memorize] record the value of the variable
[Abort] return to the planning menu

Screen management:

[Scale] modify the ratio of slide
[Origin] adjust the original position of the coordinate
[ViewDir] modify the viewpoint (VX, VY, VZ)
[Shading] choose the shading color
[Light] set up the light source
[Zoom W] set up a local window
[Last W] return to the last window
[End W] return to the first window
[AxisOn/AxisOff] display or remove the coordinates
[EasyOn/EasyOff] display or remove the wire images
[Clr Scr] clear the screen and set up the background
[Redraw] re-show the assembled model
[Return] return to the assembling menu

5.15 Conclusions

The integrated distributed intelligent simulating environment is a more powerful and user-friendly simulation facility than the conventional simulation software. In this chapter, the integration of various graphic functions and model-based inference has been discussed. IDISE has been used in industrial product design and manufacture (such as robot systems and mechanical design). It can also be used for engineering education.

6

Gear integrated distributed intelligent manufacturing system

6.1 Introduction

This chapter discusses the integration of computer-aided design, manufacturing and testing for gears using the integrated distributed intelligent system. The integration architecture, principles and implementation for computer-integrated manufacturing systems (CIMS) are presented. An integrated distributed intelligent system for gear manufacturing has been developed, which has a parallel hierarchical structure in general, and a meta-system as its kernel to manage and control the selection, communication, coordination and operation of the subsystems. A gear integrated distributed intelligent manufacturing system (GIDIMS) is implemented to demonstrate the feasibility and capability of the integrated distributed intelligent systems.

Computer-integrated manufacturing systems (CIMS) are generally acknowledged to be the next phase to computer-aided design (CAD) and computer-aided manufacturing (CAM) development (Rembold *et al.*, 1985). It is the key to enhanced productivity, minimized production cost, and improved product quality. In CIMS, product design, manufacturing, testing and production management are integrated into a unified system, which is a very comprehensive and sophisticated environment. So far, many efforts have been made to investigate the applications of integrated software systems for CIMS (Rembold *et al.*, 1985; Yeomans, 1984;

Yeomans *et al.*, 1985).

GIDIMS has a general structure for building gear integrated distributed intelligent manufacturing systems based on distributed artificial intelligence (DAI). GIDIMS employs an IDIS approach and meta-system concept. It integrates existing CAD, CAM, and computer aided testing (CAT) systems of gear manufacturing, as well as intelligent systems that are developed in different language environments (Figure 6.1).

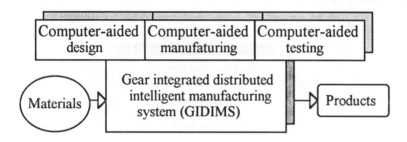

Figure 6.1 Gear integrated distributed intelligent manufacturing system.

The rest of the chapter is organized in several sections. The following section discusses the architecture of the gear integrated distributed intelligent manufacturing system. Then, the functions and structure of the meta-system and organization of meta-knowledge base in GIDIMS are discussed. What follows in the final two sections is the development of standardized communication and the optimum selection of methods.

6.2 Integration architecture

Based on its form of expression, knowledge can be categorized into two classes: analytical knowledge that is expressed in numerical models, and heuristic knowledge, which can be depicted in symbolic models. The traditional CAD, CAM and CAT techniques can only deal with analytical knowledge. These techniques can handle numerical calculation, but cannot handle symbolic reasoning. Therefore, they lack the capability of decision-making, evaluation and revision of the numerical results. To reach the stage of decision-making automation, it is necessary to incorporate AI techniques into CAD, CAM and CAT to develop intelligent CAD (ICAD), intelligent CAM (ICAM) and intelligent CAT (ICAT) as basic subsystems for CIMS. It is necessary for the CIMS software systems to provide the environment for management and integration, which can effectively combine the different intelligent systems, numerical packages, and computer graphics programs into a unified complete software platform.

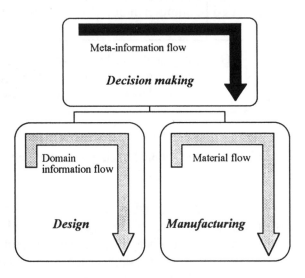

Figure 6.2 Integration of material and information flows.

In the manufacturing process of a product, an initial 'blank' becomes a product through the stages of manufacturing and testing, which are operated based on the information provided by the design stage. There are two flows in the process: material flow and information flow. The material flow is defined as all necessary working stations where the workpiece is processed, as well as the transportation of the workpiece between these stations. The information flow can be divided into two branches based on their functions: one is the domain information flow (or technical information flow), another is the meta-information flow (or decision-making information flow) as shown in Figure 6.2.

The domain information flow is defined as the processing and transforming of the information about product design, manufacturing and testing in the system. The meta-information flow is that information about integration, management and coordination among all stages. The meta-information flow controls the domain information flow and material flow, and enables them to combine and go through the system smoothly. These three flows compensate each other and coexist to form a complete system.

In such a way, an architecture of integrated distributed intelligent manufacturing system is proposed as shown in Figure 6.3. It includes a meta-system, ICAD subsystem, ICAM subsystem, and ICAT subsystem. Each component is implemented in different languages and utilized independently. The meta-system is the kernel of the system and functions as the system integrator, controller, manager and coordinator. With this configuration, GIDIMS is developed to carry out gear design, manufacturing and testing as well as fault diagnosis within one integrated distributed intelligent environment.

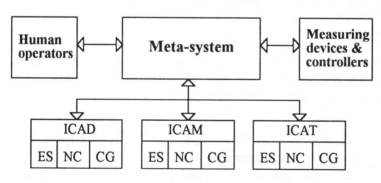

ES: Expert Systems NC: Numerical Computing   GC: Computer Graphics

Figure 6.3 Architecture of the GIDIMS.

The meta-system plays a key role in GIDIMS for system integration, management, coordination, control and communication. Written in TURBO-PROLOG™, it can take orders and tasks from users through an interface, then distribute the tasks to different subsystems and monitor the processes carried out in subsystems. It controls the whole system in many aspects.

The ICAD (intelligent CAD) subsystem engages gear design. In GIDIMS, the ICAD subsystem consists of an expert system for gear design, which is written in an expert system tool INSIGHT 2+™, and several programs written in FORTRAN™ and QUICK-BASIC™ for gear analysis and calculation. ICAD can also make mechanical drawings.

The ICAM (intelligent CAM) subsystem is in charge of process planning, programming for numerical control (NC) machines. In GIDIMS, ICAM is developed using the M.1™ expert system tool for gear process planning and programming, and using FORTRAN™ for analysis and calculation. ICAM can provide process programming for NC machines based on the information supplied by the ICAD subsystem.

The ICAT (intelligent computer-aided testing) subsystem performs the error testing on products and fault diagnosis. In GIDIMS, ICAT includes an expert system written in M.1™ for testing and diagnosis, and several programs for signal analysis and feature extraction, which are written in FORTRAN™, QUICK-BASIC™ and MACRO ASSEMBLY™ languages.

The general numerical package is a set of numerical calculation algorithms and programs for general purposes. In GIDIMS, it contains algorithms for solving differential equations and packages for statistics and optimization.

The control package for equipment is established for intelligent instruments as on-line equipment, such as programming logic controllers (PLC), numerical controllers (NC) and measuring devices. GIDIMS is not connected with the above instruments directly, and has only analog/digital (A/D) and D/A cards for data transformation and exchange.

Thus, GIDIMS coordinates numerous heterogeneous expert systems and numerical packages that are developed in several different computer languages, and integrates many functions and tasks. GIDIMS is installed and implemented in a personal computer hardware platform.

When running the software, the meta-system is invoked first and meta-information is generated to select, invoke, monitor and control the subsystems to produce domain information. The two branches of information flows control the equipment to form the material flow. Similarly, the meta-system is the center of all activities for system integration.

6.3 Functions of the meta-system in GIDIMS

In an integrated manufacturing system, from the viewpoint of a large knowledge integration environment, the meta-system can be referred to as the control mechanism of the meta-level knowledge, which was defined as the knowledge about knowledge by Davis (1987), to acquire, maintain, integrate, coordinate and utilize knowledge from different domains. The meta-system can be viewed as an expert system to control, supervise and coordinate the subsystems, as well as to solve the conflicts among them (Chapter 2). Like common expert systems, the meta-system has its own inference engine, database and knowledge base, but it does not provide the solution to any concrete problem within the specific knowledge domain; it supervises and manages the whole software system.

The meta-system of GIDIMS consists of an inference engine, database, meta-knowledge base, static blackboard and dynamic database (Cache) etc. (Figure 2.3 in Chapter 2). The static blackboard (SB) is used in the meta-system for two purposes: (1) it serves as a common station for communication between the heterogeneous subsystems implemented in different languages or tools (such as M.1™ and FORTRAN™). With SB, the meta-system executes the standardization and data transformation from subsystems; (2) it also serves as a communication transformation station for these subsystems which cannot be invoked in memory at the same time because of space shortage.

The meta-system drives the inference engine for reasoning and decision-making, based on knowledge and data in the meta-knowledge base and the database. Its functions are summarized as follows.

The meta-system manages all symbolic reasoning systems, numeric computation routines, as well as computer graphics packages in GIDIMS. It is responsible for selecting and operating all subsystems and coordinating the execution of different tasks. In GIDIMS, the meta-system can set an order to invoke subsystems or routines, which are needed for a specific task. If there is a new program needed to be added into the system, the meta-system can activate the software for editing and compiling. For example, the meta-system can use an editor program to modify a FORTRAN™ program or the knowledge base in M.1™. The meta-system also monitors the operating process within a subsystem, records the data (such as execution time), and processes data files (such as saving, making backup or deleting files).

The meta-system can solve the conflicts raised by subsystems and find the near optimal solutions. In an ICAM subsystem, if the process planning results in conflicting solutions between machining costs and an improper precision requirement from ICAD, then ICAM can bring up the problems to the meta-system. The meta-system then notifies the user and gives him/her suggestions for the solution. The user corrects conflicts and the meta-system passes the new information to ICAM. Currently, we are developing an automatic path for the meta-system to be able to find the cause of the problem and check the ICAD decision-making process, then revises design by either changing or replacing the rules related to the improper decision or by correcting the design conditions and facts. In this way, the meta-system can automatically solve conflicts that come from ICAD and ICAM. The capability of conflict optimal decision-making is extremely important for the meta-system to manage subsystems. However, it is a very complicated and difficult topic and should be further explored.

The meta-system can select an optimal method for a specific task from several available methods. In GIDIMS, some meta-knowledge about the methods of diagnosis and manufacture in subsystems is represented and can be used for method selection.

The meta-system should be able to figure out the causes and reasons for running time breakdown occurring in subsystems and then fix it automatically or notify and advise the user to fix it. In GIDIMS, the meta-system can fix a breakdown caused by lack of the necessary routines in a subsystem. When a subsystem fails to continue execution because of the absence of a predefined routine, the meta-system can seek this software routine out from other packages and invoke it to set the procedure running again. For some failures caused by other reasons, the meta-system can notify the user and advise the user of the means to fix the problem, but cannot handle them automatically.

The meta-system can communicate with the instrumentation (such as the measuring devices and the final control elements in control systems) and transform various nonstandard input–output signals into standard communication information for all subsystems. In GIDIMS, various computer languages, such as FORTRAN™, QUICK-BASIC™, MACRO ASSEMBLY™, M.1™, INSIGHT+2™ and TURBO-PROLOG™, are employed to build different subsystems. The standardization of communication among them is very significant for the integration of these subsystems. A detail discussion about this issue will be given in a later section.

The meta-system should provide convenient and efficient interface to increase productivity of an integrated manufacturing system (Su and Lam, 1990). The interface should be designed for users with various capabilities, levels and needs to ease the tasks of using the system. GIDIMS is equipped with an interface for users at two levels. For novice users, the meta-system can conduct reasoning, give suggestions and advice, make decisions and explain the process. The system thus can work for training purpose. For expert users, the meta-system allows them to retrieve, manipulate and modify the objects of knowledge bases in subsystems, as well as to be involved in the process of judging, controlling and decision-making. In this way, the system function can be greatly improved by codifying more creative knowledge and experiences from human expertise.

The meta-system can distribute knowledge into the separate expert systems to maintain and improve knowledge bases. In such a manner, the knowledge bases of subsystems can be easily modified by end users other than their original developers. The system can be self-improved by its own experience. In GIDIMS, the meta-system can obtain feedback from ICAT to test results and then add the new knowledge into the knowledge bases of ICAD and ICAM. For instance, if the testing results indicate the middle concave on gear teeth, then a rule will be added to the knowledge base of ICAM to increase the precision of gear shaving tools; if the major failure is always the plastic flow of the surface material on gear teeth, then a new rule may be inserted into the knowledge base of ICAD to increase the value of hardness, or added into the knowledge base of ICAM to change process planning for heat processing of gears.

6.4 Modularity of meta-knowledge in GIDIMS

By different functions, knowledge can be categorized into 'domain' knowledge and 'meta-level' knowledge. The former is usually defined as facts, law, formulae, heuristics and rules in a particular domain of knowledge about specific problems, whereas the latter is defined as knowledge about domain knowledge and can be used to manage, control and utilize domain knowledge. There are some differences between the two categories of knowledge. For example, meta-knowledge possesses diversity, covering broader areas in contents and varying considerably in nature; also, it has a fuzzy property, more uncertainty than determination. Hence, the representation of meta-knowledge and the structure of meta-knowledge base (MKB) should have their own characteristics.

In GIDIMS, the meta-knowledge is represented by combinations of production rules and frames. The meta-knowledge base is constructed in modules, shown in Figure 6.4. Serving as a manager, the MKB-M module is the only one connected to the inference engine, and passes its reasoning commands to other modules. The subsystem management module (SS-M) possesses the knowledge about subsystems, such as the language environment, the knowledge representation techniques, the inference mechanisms, the functions, and the control strategies. Through this knowledge module, the meta-system can communicate with different subsystems and choose the right one to perform a certain task.

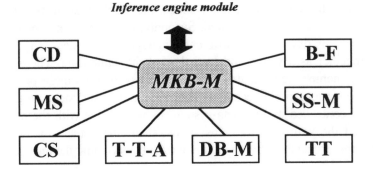

CD: Conflict Decision-making   MS: Method Optimum Selecting
CS: Communication StandardizationT-T-A: Task-Taking-Allocating
DB-M: Data Base Management   SS-M: Subsystem Management
B-F: Breakdown Fixing   TT: Technical Training
MKB-M: Meta Knowledge Base Management

Figure 6.4 Configuration of meta-knowledge base.

In GIDIMS, the database management module (DB-M) is responsible for communication between the meta-system, written in TURBO-PROLOG™, and the database, DBASE III™. The knowledge about the data structure and functions of DBASE III™ and TURBO-PROLOG™ are stored in the module for data transformation and data file processing. The modules in the meta-knowledge base are independent, hence it is easy to modify, delete and add to any module without bothering the others. All modules connect to each other through the MKB-M module, and can share knowledge with each other. This property makes the meta-knowledge base more flexible and efficient.

6.5 Standardization of communication in GIDIMS

As a large integrated knowledge environment, GIDIMS faces two barricades for implementation on a single computer. First of all, due to the limitation of the on-board memory space, the whole system may not be able to be loaded at the same time. Secondly, due to the heterogeneity of the subsystems developed in different languages, the system may not be linked together as a whole and loaded in at the same time. Therefore, a technique, namely covered-structure, is adopted to solve these problems. The basic strategy of the technique is to invoke the subsystems in turn by an order, controlled by the meta-system, in accordance with the memory space and the nature of the subsystems, then find some way to allow information exchange between the invoked subsystem and the ones at rest.

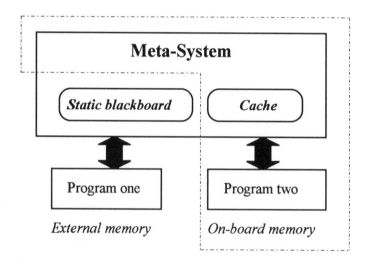

Figure 6.5 Communication between meta-system and subsystems.

GIDIMS adopts the covered-structure technique and solves the communication problem in the following manner. Shown in Figure 6.5, there are two ways for communication, namely, direct communication and indirect communication. The direct communication method is employed for those programs which are used frequently and require quick access to the meta-system in order to transform data quickly. Under the conditions of enough memory and compatibility of the languages, the programs may be compiled and linked with the meta-system and reside in the memory for the whole process. In this way, the meta-system can communicate with the program directly. For example, a real-time control system for machine tool needs to get information about states of the machine within a second, from sampling signals of vibration or noise on the machine to obtaining the results calculated by a statistics program. The communications among the meta-system, numerical packages, measuring devices and the control systems have to be very fast. In such situation, the statistics program and data acquisition software should communicate with the meta-system directly to guarantee the control systems obtaining the necessary data in time. Through standardization, the meta-system can accept data from the programs linked with it, fast communication can thus be achieved.

The indirect communication is realized through a 'transformation station', namely a static blackboard (SB). The meta-system can invoke some executable programs by using the predicate SYSTEM(" ") in TURBO-PROLOG™. The results from the running programs are then sent to SB, and processed there into the standard forms for calling by other subsystems through the meta-system. For example, the diagnosis expert system of ICAT in GIDIMS communicates with other parts of the system in such a manner. The procedure of indirect communication is illustrated in Figure 6.6 and takes several steps, as follows:

(1) Based on the meta-knowledge, the inference engine sends program P1 the instruction. Program P2 obtains data from the instruction, then passes the language type of P2 to the instruction.

(2) The inference engine invokes P1, and P1 starts to run.

(3) The running results of P1 are written in SB with a file name, such as 1.DAT. The data file informs the meta-system.

(4) The meta-system identifies the data file by the name 1.DAT and standardizes it, then saves the results in S1.DAT.

(5) The meta-system sends S1.DAT to program P2.

Indirect communication is complicated and slow. In GIDIMS, data to be written into SB should be prepared in a certain structure. An example is presented:

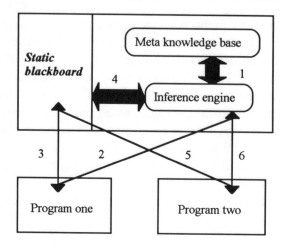

Figure 6.6 Indirect communication procedure.

Transformation Symbol: "FM"
Variable Name: "Conclusion"
Logical Relation: "Equal to"
Value: "1024.6"
Certainty Factor: "cf90"
Data Feature: " . "

The data structure includes six segments. The Transformation Symbol indicates the two languages of the programs between which the data is going to be transformed. For instance, "FM" means the data will be transformed from a program written in FORTRAN™ to a program written in M.1™. The 'Logical Relation' stands for the relationship between the variable name and its value. For example, it may be 'greater than', 'less than', 'equal to' or 'belong to', etc. The 'Certainty Factor' (CF) stands for the degree of facts being true. In the above example, the certainty of the value 'Conclusion = 1024.6' is '90%'. The term 'Data Feature' refers to a special data structure that some languages may require (for example, M.1™ requires a dot '.' at the end of all data).

The procedure of standardizing the data in SB takes two steps. First, the meta-system identifies the two languages from the transformation symbol. Then, based on the meta-knowledge about languages, the data are processed into standard form acceptable to the receiver.

A commercial database, DBASE III™, is adopted in GIDIMS, thus the implementation of communication between TURBO-PROLOG™ and DBASE III™ is an important content in the database management module (DB-M) of the meta-knowledge base. The following is a simple description of transformation between TURBO-PROLOG™ and DBASE III™.

Since TURBO-PROLOG™ can only recognize facts in symbolic sentences, whereas the database can only store records, the key issue for communication between them is how to convert the records in the database into symbolic form acceptable to TURBO-PROLOG™. The predicate is designed for this purpose:

Database_To_Prolog (Fact_name, Database_files_name, Begin, Amount)

Here, Fact_name stands for the fact name having been transformed to TURBO-PROLOG™. Data_files_name denotes the file name in the database to be transformed. 'Begin' stands for the first record number to be transformed to TURBO-PROLOG™. 'Amount' denotes the number of records transformed to TURBO-PROLOG™. The transforming process using TURBO-PROLOG™ is given as follows:

- Identifying the file name in the database in which the data are to be transformed, opening temporary files for transforming process, and reading in the starting address and record length.

- Reading attribute values of the records, and establishing an attribute table for these values by using the retracing and list processing function provided by TURBO-PROLOG™.

- Determining the position of first transforming record according to initial recording mark 'Begin'.

- Reading all records by the order given in the attribute table, transforming the records to facts by a series of predicate operations. The facts are saved in the temporary files.

- Reading facts from the temporary files to TURBO-PROLOG™

system. Then the facts are further saved into the dynamic database of the meta-system.

The above discussion only covers communication between the meta-system, database, expert systems and numerical programs. The communication between the meta-system and equipment, such as the machining center, needs further investigation in the future.

6.6 Optimal selection of methods

When there are several methods in the system to accomplish a number of tasks, it is important to select the most appropriate one for each specific task in order to raise efficiency and robustness. For instance, in GIDIMS, there are four methods for gear fault diagnosis and three methods for gear shaving, and each of them has its own characteristics. Selecting the best one for a particular task involves symbolic manipulation and reasoning, so the meta-system should adopt the expert system technique to solve the problem. In the meta knowledge base of GIDIMS, there is a module to provide the knowledge about those methods which can be employed to deal with these kind of tasks.

Finally, we discuss the way to choose the optimum method for gear fault diagnosis. Four methods applicable to gear fault diagnosis are listed below:

- spectrum analysis method, written in FORTRAN™ and MACRO ASSEMBLY™ for numerical calculation (nc),

- time series method, written in QUICK-BASIC™ (nc),

- major component method, written in QUICK-BASIC™ (nc), and

- parameter statistics method, written in FORTRAN™ and MACRO ASSEMBLY™ (nc).

The symbolic processing for the four methods is developed with the expert system tool M.1™. To select the optimal method, an optimization model should be established to evaluate the methods. The following four factors are considered for the evaluation function $S_F(X)$:

- diagnosis precision of the method, denoted by D_p,
- diagnosis speed of the method, denoted by D_s,
- expressive effects of the method, represented by E_x, and
- convenience of the method in operating, appraised by C_o.

Thus, the evaluation function is established as follows

$$S_F(D_p, D_s, E_x, C_o) = k_1{}^*D_p + k_2{}^*D_s + k_3{}^*E_x + k_4{}^*C_o + k_5 \qquad (6.1)$$

where,

D_p, D_s, E_x, C_o = appraising factors,

K_1, k_2, k_3, k_4 = weighting coefficients, and

k_5 = revising coefficient.

Some constraints may be incorporated in the model for specific diagnosis tasks. For example, to diagnose the original fault in gears, precision should be defined to a specific value and speed is less important. For on-line diagnosis, high speed is required and precision should be given a lower bound. The best method should have the highest value of $S_F(X)$ that also satisfies the constraints.

In equation (6.1), the parameters k_i ($i = 1, \dots, 5$), D_p, D_s, E_x, C_o need to be determined by experience and knowledge of human expertise. In the system, this heuristic is represented in the meta-knowledge base, and based on it, symbolic reasoning is carried out to determine the optimal solution.

6.7 Conclusions

The principal features of GIDIMS are summarized below:

The hierarchical structure of an integrated distributed intelligent system is suitable for integrated manufacturing systems for its flexibility and easy management. The meta-system is adopted as the kernel of the system to integrate, manage, control and utilize subsystems. Although the meta-system is similar to ordinary expert systems, it has a unique structure and functions. The use of a modular structure for the meta-knowledge base is convenient for managing the meta knowledge and efficient for inferring.

System communication is carried out by direct and indirect means. The indirect communication adopts a static blackboard as transformation station, which is an effective communication method suitable for hierarchical systems.

Optimal selection of methods should be done with a combination of expert system techniques and numerical optimization methods. A proper optimization model with an evaluation function should be established based on heuristics.

7

Intelligent system for process startup automation

7.1 Introduction

An integrated distributed intelligent system is developed to assist the process startup automation in a section of a refinery. It is developed in conjunction with an industrial partner. This intelligent system codifies the knowledge from industrial process operators with many years of experience. The project aims at investigating automation for the process startup operation by using an integrated distributed intelligent system concept as well as capturing the important expertise knowledge. Testing and utilizing the intelligent system within the plant indicates that the system can provide operational advice at the level of human experts. Another element in the success is the automation of 'look-up' in the written manual. Hypertext is used to make access to the manual rapid and efficient. The hypertext system is integrated with the knowledge base to provide intelligent selection of appropriate manual sections. An important result of this research is the development of a new conceptual process model. A model to represent the chemical and mechanical states of the unit for startup is constructed. The model is characterized by the mechanical states, modified by the operators, to manipulate the chemical states. Then, the chemical states are driven towards the goal states: the normal operation of the unit. This methodology provides a new approach to improve the industrial manufacturing environment using the newest AI technology.

The growing complexity of industrial processes and the need for higher efficiency, greater flexibility, better product quality, and lower cost have changed the face of industrial practice. It has been widely recognized that quality control and process automation are the key elements to make modern industries stay competitive internationally. Recently, the chemical companies have begun to recognize the importance of process automation in order to function successfully as a manufacturing facility (Astrom, 1985).

The technological advances in automatic control have addressed the research interests of intelligent control, which encompasses the theory and applications of both artificial intelligence (AI) and automatic control. Intelligent process control is interdisciplinary in nature, and allows the application of knowledge from systems engineering and computer science to chemical processes. It is developed for implementing process automation, improving product quality, enhancing industrial productivity, preventing environmental and hazardous risks, and ensuring operational safety. Its objectives are to enhance product quality through automation, and to improve production efficiency through AI techniques.

As we know, the main objective of most chemical processes is to convert raw materials into high-quality products at the lowest cost and with low impact on the environment. Therefore, a chemical plant that is well designed, well controlled and well operated, should meet the above requirements.

The safe operation of a chemical process depends on the human operator and the real-time process control system, and both of them must work together. If the processes can be well described by mathematical models, then they can be controlled very well by real-time control systems, thus few human skills and expertise are usually required for the desired operations. However, many chemical industrial processes are very complex in nature, and the development of mathematical models for process control is so difficult that the performance of real-time control systems is poor. In such a case, only the most experienced and skilled operators can handle the operation.

In the future computer-integrated manufacturing environment, process control systems will not allow many interruptions by human operators under special or dangerous situations. There exist four operation stages for chemical manufacturing processes. They are: startup, normal operation, emergency situation and shutdown. The majority of process automation research and development focuses on normal operation, and this has been very well studied, and run under fully automatic control. However, the other three stages are the most dangerous but are still operated manually, and suffer from a lack of sufficient attention from academic or industrial research. Today, it is very necessary that industrial practice provides automation technology to improve such labor-intensive operations. There is no production outcome during process startup. A safe and quick startup of a chemical process reduces production costs, and brings valuable economic benefits to a company.

The following difficulties in the operation of a current chemical process startup regime are noted:

- Process startup is the most dangerous operation. The majority of process accidents occur during process startup. It is operated manually, inefficiently and unsafely.

- There are many ill-structured problems that are difficult to solve by conventional modeling plus numerical control methods.

- Since there are very few opportunities for startup/shutdown operations in the continuous manufacturing processes, most human operators are unfamiliar with such operations. There are only a few experienced operators who can safely carry out startup process operations.

- There is little process operation information available and there exists little interaction between the operators and the controlled process during process startup operations.

- The valuable private knowledge about the startup operation cannot be accumulated, then transferred for public use in the company.

Most AI applications in chemical engineering have been for process control or design (Stephanopoulos and Davis, 1990). Process startup automation and process operation support are other areas where intelligent systems could be of value. However, very little previous research has been done in these areas.

An intelligent system, namely IPSAS (intelligent process startup automation system), has been developed through a cooperative research effort between the Intelligence Engineering Laboratory at the University of Alberta and an industrial partner. The objective of this research is to develop an intelligent system to automate the process startup operation by providing proper and safe startup procedures. Many injuries and fatalities in the chemical process industries have been due to human errors during the operation of a plant (Kletz, 1988). Proper procedures must be followed during plant operation in order to reduce incidents (Croce *et al.*, 1988). With the use of computer control for plants, some increase in safety is achieved but other hazards arise. This is due to the computer's inability to perform as a human operator. It lacks the intelligence and expertise of the operation experts (Pearson and Brazendale, 1988). Thus, our project concentrates on the development of an intelligent system to select proper and safe startup procedures.

7.2 Process startup operation

The process unit under consideration is a gas oil treater used to produce a low-pour oil. Details of the process have been left out in order to protect the proprietary information of the plant. The two main process streams contain oil and hydrogen, each under high pressure and temperature. These streams, as well as the fuel and high pressure steam lines, pose the greatest potential for danger in operating the unit. A simplified scheme of the process is shown in Figure 7.1.

The feed oil comes from another area of the plant and is passed through a preheat train. This consists of a series of heat exchangers and one fuel gas furnace. Feed hydrogen originates from a steam reformer located at the plant. The feed hydrogen is combined with recycle hydrogen and passed through a compressor.

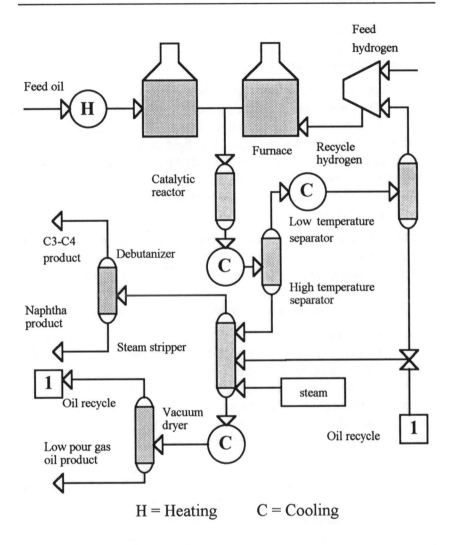

H = Heating   C = Cooling

Figure 7.1 Simplified scheme of the process.

The combined stream is heated in another fuel gas furnace and then combined with the heated feed oil. The catalytic reactor breaks down the large oil molecules as well as removing sulphur as hydrogen sulphide. The outlet from the reactor is cooled and then passed to a separator train. The first separator passes the condensate to the purification section. The 'overheads' are cooled further and passed to another separator.

The 'bottoms' of the second separator are also sent to the purification section. The overheads are stripped of hydrogen sulphide and then returned to compression as recycle hydrogen. The initial action of the purification stage is steam stripping. The light ends are driven out of the stripper and sent to a debutanizer. The bottoms of the steam stripper are dehydrated in a vacuum dryer. The dried bottoms are sent to tankage as the main low-pour product. The wet overheads are returned to the steam stripper as recycle oil.

There are two modes of operation for this unit. During the winter, the unit operates in the manner described above. During the summer, a slightly different product is produced. The debutanizer is mothballed and not used. These two different modes of operation complicate operating the unit. These modes will be referred to as winter and summer operation.

Safe startup requires the correct procedures and good leadership from the senior operators. This unit is selected for prototype development because it has a good set of written procedures as well as several on-site experts. The operators are also motivated in developing new technology to improve operation safety. Furthermore, the unit has a scheduled shutdown twice a year. This regular shutdown and startup schedule ensures the continuous use of a startup automation system.

The startup procedures are made up of seven main phases. In each of these startup phases there is a set of procedures to be carried out. For each of the procedures there is a step-by-step set of instructions. Table 7.1 summarizes the startup phases and their procedures. In total, there are 40 procedures grouped into 7 phases. The procedures performed depend on the mode of operation and the type of shutdown. Shutdowns can be either total or partial. Total shutdowns are for major maintenance and switching modes of operation. Two total shutdowns are scheduled each year. Partial shutdowns occur when minor repairs are needed. Minor repairs include valve replacements, repairing leaks, cleaning heat exchangers, and so on.

These startup phases do not include the base level utilities that are commissioned during an initial startup and left on-line continuously. These extra phases are very uncommon during startups other than the initial startup for a new unit. Utilities that are included in these phases are the closed drain header, steam header, instrument air, etc.

Table 7.1 Summary of the startup procedures

1) General preparation	4) Establish recirculation
a) check fire extinguishers b) check safety equipment	a) back in naphtha into debutanizer b) back in oil into separators
2) Oil circuit preparation, air free and tightness test (AF&TT)	c) oil fill the feed preheat circuit d) pressure test preheat with oil e) commission oil & vacuum circuits
a) oil free circuit	5) Catalyst preparation
b) steam stripper and overheads c) stripper bottoms	a) caustic treating b) presulphiding
d) debutanizer	6) Hydrogen circuit preparation
e) vacuum dryer f) vacuum overhead g) vacuum bottoms h) recycle oil separator i) purge gas knock-out drum j) fuel gas circuit	a) start compressors b) warm up hydrogen circuit with nitrogen c) pressure test with nitrogen d) flush with hydrogen e) establish hydrogen recycle
3) Pre-startup checklists	**7) Initiate reacting and slowly ramp full capacity**
a) fire hydrants	
b) furnace dryout c) hydrogen circuits AF&TT d) check blanks e) car seals f) safety valves g) emergency valves h) alarm plugs i) shutdown trip switches j) analyzers	a) warm both oil and hydrogen circuits b) commission lean MEA c) pump feed oil into unit d) swing feed into reactor e) establish debutanizer reflux f) slowly ramp to full capacity

7.3 Chemical/mechanical states

Before the startup procedures can be recommended, the state of the unit must be identified. The state determines how the startup procedures are affected. One of the benefits of IPSAS construction is that the domain knowledge becomes better organized and easier understood. During organization of the knowledge base for IPSAS, a new model for the process operation has resulted. The model that is used to describe the states of the physical plant has three conceptual components: network connection, chemical states and mechanical states.

The chemical states of a unit describe chemical processing characteristics like flow rates, temperatures, pressures, types of chemicals, degree of reaction, etc. The chemical states are the parameters that must be shifted from their current states to the goal states: the normal operations specified in process design. In IPSAS, the chemical states are used to identify the conditions for the main process streams and operations. For example, the hydrogen stream needs to be maintained at a certain temperature and pressure before it is mixed with the oil stream and sent to the reactor. Similarly, the oil stream must also have a certain threshold temperature and pressure. These conditions must be met before a chemical reaction occurs otherwise the reactor catalyst may become poisoned.

The mechanical states describe the physical location and positioning of the equipment as well as their conditions, such as valve settings, state of blanks (in or out), vessel temperatures/pressures, location of bleed valves, elbows and other constrictions, etc. Usually, process simulation relies entirely on the chemical states of the unit. In modeling the startup problem, this is not sufficient.

The startup procedures are operated directly on the mechanical states to achieve changes in the chemical state. To drive the hydrogen and oil streams to their threshold conditions, bypasses and recycle routes must be set up. This is done by closing valves on the normal piping routes and opening others that redirect the hydrogen and oil away from the mixing point and reactor. Meanwhile, steam and fuel gas are supplied to provide heat to these streams. Heat exchangers and furnaces provide the necessary heat. Compressors and pumps add pressure to the streams. These actions represent modifications to the mechanical state and indirectly affect the chemical states.

Network connection is used to present the chemical and mechanical connections among all pieces of equipment. The connection description provides a means to determine the interdependence between the equipment. The network connection concept is implemented by grouping individual equipment into circuits. These circuits represent sub-units of the plant section. As shown in Table 7.2, performing a specific task, each circuit consists of different equipment. These circuits are described in detail later.

Table 7.2 List of the circuits in the plant unit

Circuit name	Equipment within the circuit
Preheat circuit	E-01, E-02, E-03, E-04, E-07, E-08, E-09, F-01, Piping
Stripper and overheads circuit	D-04, E-10, E-11, E-12, E-18, E-19, T-01, piping
Stripper bottoms circuit	D-06, E-02, E-03, E-04, E-15, E-16, piping
Debutanizer circuit	D-05, E-15, E-20, E-21, E-22, T-02, piping
Vacuum dryer circuit	D-07, E-23, T-03, piping
Vacuum overheads circuit	C-05, C-06, D-14, E-24, piping
Vacuum bottoms circuit	E-17, piping
Recycle oil separator circuit	D-07, piping
Purge gas KO drum circuit	D-12, piping
Fuel gas circuit	D-08, piping
Hydrogen gas circuit	C-01, C-02, D-03, D-09, D-10, D-11, E-05, E-06, E-07, E-08, E-09, E-10, E-11, F-02, R-01, T-04, piping
footnote: R = reactor, T = Tower	D = drum, C = compressor, F = furnace E = heat exchanger

7.4 Knowledge acquisition and representation

Knowledge acquisition is the most difficult task during developing intelligent systems. In this project, it is facilitated by several factors. First, there is an extensive set of written procedures available for the unit. These manuals are written by a member of the original design team for this plant section. The procedures have been modified and refined over 6 years of implementation. Thus, they form the basis for the knowledge of IPSAS.

There are also several experts available for consultation. The main expert is a senior operator with many years of experience. He is a member of the original design team and is the most knowledgeable person about this section in the refinery. As the original author of the startup procedures manual and the most experienced operator for this plant section, his input to IPSAS is fundamental. Furthermore, this expert is nearing retirement so it is paramount to capture and retain his private knowledge and keep it available for public use within the plant.

Finally, the developers of IPSAS are trained in chemical engineering and intelligence engineering. Through the written procedures, expert consultation and first-hand experience, we are able to acquire the necessary knowledge to build IPSAS. Having a background in chemical engineering greatly assisted the construction and organization of the knowledge base.

There are two levels of knowledge represented in IPSAS. The first is a plant model based on the chemical/mechanical state model. The second level of knowledge is the relationship between the plant model and the startup procedures.

The plant model is constructed from 11 sets of operation sequence of equipment where each set represents one circuit within the plant unit. Grouping the unit into circuits provides information about the network connection of the equipment. The circuits are based on recommendations from the design and operation teams. As shown in Table 7.2, each circuit is made up of many pieces of equipment. The circuits are physically separated from each other by block valves. The circuits are modeled by sets of equipment that match the actual plant equipment. Some heat exchangers are included in two sets. This represents the different circuit flows passing on the shell or tube side of the exchanger. For example, feed oil is passed through the shell side while the reactor outlet is passed through the tube side. The streams transfer heat but do not mix. This heat exchanger is considered as part of both the feed oil preheat circuit and reactor effluent circuit.

For IPSAS, only major equipment is included in the sets. The more detailed case involving blank locations and minor equipment is not considered. A distinction is made between preparation procedures and startup procedures. Preparation procedures, including blank placement and removal, equipment opening and closing as well as repair procedures, and preparation of equipment for the standard startup procedures. The startup procedures are performed to get the unit from its current prepared state to the normal operation state. The level of detail implemented in this model is sufficient for the startup advisory system. If preparation procedures are to be included, then more detailed modeling should be implemented.

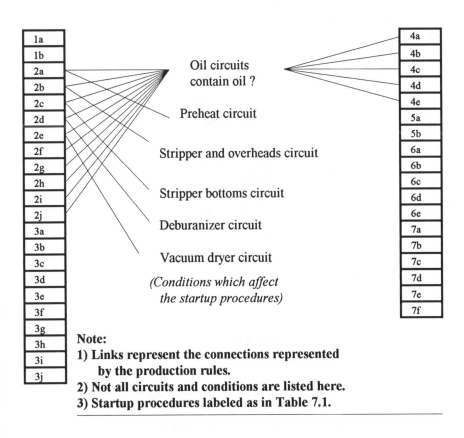

Figure 7.2 Illustration of the knowledge structure.

The second level of knowledge describes the relationship between the plant model and the startup procedures. These relationships are represented through production rules. There are two types of production rules. The first involves the effects of the operating conditions on the startup procedures. The mode of operation, types of process fluid still in the unit and the type of shutdown are represented in the operating condition rules. The other set of rules covers the effects of opened equipment. A list of opened equipment is built from the user's input. This list is used to determine which circuits are affected. The rules relate circuits with opened equipment to a subset of startup procedures. These subsets are composed of procedures from several startup phases. Figure 7.2 illustrates the representation format for the knowledge.

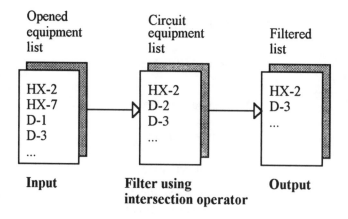

Figure 7.3 Illustration of rule filter.

The rules governing the effects of the opened equipment are implemented using a technique similar to the concept intersection method (Winston, 1975). This technique can be viewed as a filtering process. Each rule tests the opened equipment list within a given circuit. The filtering action is illustrated by Figure 7.3. If information succeeds in passing through the filter, then the rule fires and modifies the startup procedures. The filtering process is performed through an intersection operator. The opened equipment list is intersected with a circuit equipment list to determine whether the equipment within the circuit has been opened. Beyond the implementation details, the rules can be viewed with the following syntax:

IF (equipment in circuit X has been opened)

THEN (modify startup procedures to bring this circuit to normal operation).

7.5 Hypertext technique

Hypertext is a technique of linking a keyword or phrase to text that further explains the original keyword or phrase (Gessner, 1990). This technique is mainly used as an information retrieval system. An example of hypertext is an encyclopedia. An encyclopedia contains a large volume of information but is organized into a functional form. The information is grouped into topics according to a short keyword or phrase. At the end of each article is a list of the related topics and the keywords to search under. This process of retrieving more information from related sources is fundamental to accessing large amounts of information. Hypertext is used to accomplish this task.

In IPSAS, hypertext is used to obtain the startup procedures and their step-by-step instructions. The startup manuals are entered into the computer using a word processor. The text files are then linked through hypertext. Hypertext allows operators to select information without knowing the page numbers or searching through unrelated text.

The development of new intelligent systems is changing rapidly both in ease of construction and time required due to intelligent systems building tools. In the selection of intelligent systems developing tools, commercial tools are the first priority. The reason is very straightforward since a sophisticated, well-endowed development and run-time environment can save us a great deal of work generating basic facilities, including reporting, debugging, graphics, database, statistical packages, and other specialized functions.

7.6 Implementation

In the implementation of intelligent systems for operation and control of industrial processes, three practical phases are suggested (Rao, 1992):

Phase 1: Off-line system

The objective at this phase is to organize the intelligent system, to codify the knowledge, and to evaluate the intelligent system in an off-line environment, such as design and training as well as off-line supervisory control. A system at this phase is basically a simulation where the entries are performed by the user over via a keyboard. Such a system is excellent for training purposes because it is a highly interactive system. Most academic intelligent systems are developed at this phase. Its main function is to provide decision support to a human operator, based on process operation information.

Phase 2: On-line supervisory system

The objective of this phase is to evaluate the intelligent system interface with the human operator, process computer systems and hardware instrumentation. The intelligent system is physically connected to the actual process. The process information is fully or partially processed by the intelligent system, and reports and suggestions are generated.

Phase 3: On-line closed loop system

A closed loop intelligent system directly reads the inputs from the process and sends its outputs (actions) back to the process. In most cases, the objective of these systems is to reduce the operators' routine tasks and improve operation safety and efficiency.

IPSAS is implemented at phases 1 and 2 where the data are entered by the user using a keyboard. Currently, IPSAS is an off-line system but connected to the process control mainframe computers. However, it can provide on-line supervisory control that helps the operators to control the process startup operation.

This prototype is developed as the initial stage in the overall development of an on-line, real-time intelligent system for process startup operation automation.

Being different from other applications in this book, the implementation of IPSAS relies on using commercial software systems. Such a strategy aims at investigating how to integrate and utilize existing facilities in the plant and stimulating industrial partners to use the newest technology by protecting their investments.

Based on the concept of the integrated distributed intelligent system, IPSAS is developed on a DOS™-based microcomputer running Microsoft Windows™ (MSW). Its major functions are:

- to provide decision support during the startup of the plant,
- to provide a direct connection to the process control computer, and
- to allow viewing the process and instrumentation diagrams (P&IDs).

This prototype system is built using an expert system shell called KnowledgePro™ for Microsoft Windows™, which provides immediate access to a graphical user interface. If IPSAS only functions as decision support, it and the computer would sit idle most of the time. Plant shutdowns and startups happen infrequently and IPSAS would be useful only during these times. Additional use would come from training new operators, but this would also be infrequent. To obtain real-time process data and increase the daily usefulness of the system, a direct connection to the process control computer is provided. This allows using the microcomputer as one of the standard terminals for controlling the plant. Furthermore, providing a means of viewing the P&IDs on screen is found to be a welcome feature.

The main menu system is used to select the major functions. These functions can be seen on the screen as shown in Figure 7.4. The process control computer is connected using Attachmate's Extral for MSW. It acts like a terminal session that could be iconized and brought back just like any other MSW application. To run the session, however, we also need a hardware card to accomplish the physical connection.

The viewing of the P&IDs is performed by shelling out of MSW to SirlinView's AutoVIEW. AutoVIEW is a non-MSW AutoCAD™ viewing program. This package provides the ability to display and view AutoCAD™ drawing files without the overhead needed for editing functions. An important feature of AutoVIEW is the ability to post notes to the drawing. These notes can be viewed quickly and easily by clicking on them with a mouse. This allows posting of the desired modifications to the P&IDs, making notes about special operating conditions as well as equipment safety information.

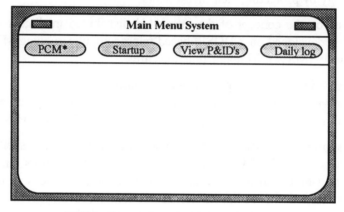

* PCM = Process Control Mainframe Connection

Figure 7.4 Main menu system.

The Daily log function is used to either view past log entries or create new ones. The daily logs are saved consecutively to a default file that could be emptied once a month. The Startup function calls up a new window that contains another menu system as well as a hypertext system.

The Startup menu system is shown in Figure 7.5. This menu system provides access to functions that supplement a normal startup sequence. Shutdown state selection is used to reenter the shutdown state in the case that one of the questions is originally answered incorrectly. Viewing of P&IDs is identical to the main menu system. Each startup session is saved in a file. These files contain a summary of the actions taken by the user. It begins with a summary of the shutdown status and follows with a time history of completed startup phases. By using the Add log function, the user can enter special notes to the startup file. These extra notes could describe difficulties that are encountered and how they are overcome so that future startups can easily obtain access to this information.

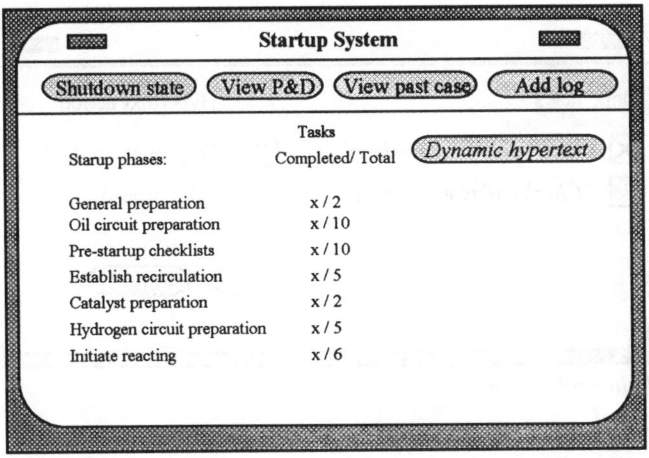

Italics = Hypertext

Figure 7.5 Startup menu and hypertext system.

The main interaction with the program during startup is through a hypertext screen that contains a summary of the present state of the startup. There are two areas where hypertext triggers are active (see Figure 7.5). On the left of the screen is a list of the main startup phases. When one of the phase titles is selected by the mouse, a window with the associated procedures for that phase is brought up (see Figure 7.6). Beside each procedure is a check box indicating whether or not the procedure is complete. Furthermore, each of the displayed procedures is also a hypertext trigger. If a procedure is selected, the step-by-step instructions are brought up. The user can choose whether the instructions will go to the screen or a printer. The other hypertext section in the startup window is a button. It dynamically changes functions according to the next step required in the startup sequence. By using the dynamic button, an operator can step through the required sequence using one function. This design is found to be the simplest and most direct approach to the user interface problem.

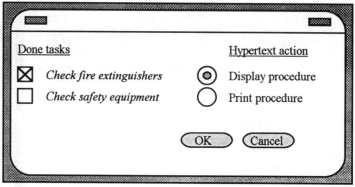

Italics = Hypertext

Figure 7.6 Phase and procedure hypertext system.

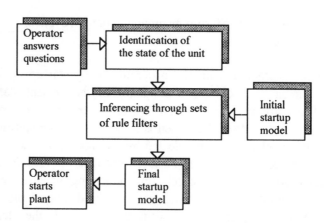

Figure 7.7 Knowledge processing overview.

Figure 7.7 summarizes the overall process of startup operation support. The operator responds to several questions posed by IPSAS. The answers to these questions are used to determine the shutdown state of the unit. Initially, the system assumes that the unit is completely shutdown and all procedures require to be gone through. The shutdown state and initial startup model are passed into the inference section. As the rules in the knowledge base fire, modifications are made to the startup model. When the inference is complete, the final startup model is produced. The results are returned to the operator as recommendations on the startup of the unit.

7.7 Case study

IPSAS has been successfully used in the plant on several occasions. It has been fully evaluated in the plant. The following example comes from the actual testing and applications of IPSAS.

During the summer of 1990, a leak had developed in a pipeline within the stripper and overheads circuit. The unit was partially shutdown to repair the leak and to replace a couple of valves on one of the hydrogen compressors at the same time. The input to IPSAS was the following:

- remaining in summer operation
- oil circuits were not drained
- hydrogen circuits were not drained
- compressor C-02 was opened
- piping in the stripper and overheads circuit was opened

IPSAS then made the following recommendation for the startup sequence:

(a) General preparations phase

- No action to be taken in this phase

(b) Oil circuit preparation phase

- Steam stripper and overhead AF&TT procedure

(c) Pre-startup checklists phase

- Hydrogen circuits AF&TT procedure

- Check blanks procedure

(d) Establish recirculation phase

- Commission oil and vacuum circuits procedure

(e) Catalyst preparation phase

- No action to be taken in this phase

(f) Hydrogen circuit preparation phase

- Establish hydrogen recycle procedure

(g) Initiate reacting and slowly ramp to full capacity phase

- Warm both oil and hydrogen circuits procedure

- Commission lean MEA procedure

- Pump feed oil into unit procedure

- Swing feed into reactor procedure

- Slowly ramp to full capacity procedure

Only 10 of the 40 startup procedures were needed to startup the unit. This is typical for partial shutdowns. The actual startup sequence followed this recommendation for all procedures except one. The deviation between the recommended and actual startup was the steam stripper and overhead AF&TT procedure (b). The operators performed a tightness test only on the section of pipe where the leak occurred. Based on IPSAS's recommendation, a contact engineer suggested that the full circuit test would have been the safer and more thorough approach. The operator in charge, however, was very experienced (the knowledge source of IPSAS is mainly based on his experience) and had complete confidence in the success of the shortcut taken, and thus rejected this advice. The contact engineer then pointed out that another similar shortcut led to a loss in a vacuum section that lasted for weeks. The loss of vacuum section was finally traced to an open bleed valve. A full circuit tightness test was used to identify potential problems of this nature. This was a classic case of conflict between experts. Both were correct but one approach may have more advantages than another. IPSAS chose the approach that provided the safest and most thorough sequence. The operators then had the prerogative to take shortcuts that they felt were valid.

The startup problem is qualitative in nature. It also makes engineering analysis an invalid approach for evaluation. The startup problem has many different solutions, as shown in the above example, where IPSAS and the operator were in conflict over the air-free and tightness test. Both approaches were correct. However, IPSAS's recommendation was more thorough. The shortcut proposed by the operator could lead to further problems like the vacuum leak previously described. In general, IPSAS provides the safest operation recommendation.

As the private knowledge of human beings is successfully codified into computer programs, it always behaves with robustness. However, an experienced expert can often make mistakes due to human factors such as forgetfulness, being confused, getting nervous when facing an emergency, and so on. This case is a good example of such factors.

One shortcoming of IPSAS is that primarily one expert's knowledge was incorporated. In fact, multiple experts' knowledge should be incorporated to provide a more flexible system. The AF&TT shortcut could be implemented safely if a specific set of checks were made prior to startup. However, this project was aimed at investigating process startup automation and identifying the potential use of integrated distributed intelligent system within an industrial setting. Concentrating on capturing the knowledge from only one expert helped to accelerate the development of this system. Future development will be aimed at capturing other experts' knowledge.

Another shortcoming is the lack of temporal reasoning that would provide time analysis to the recommendations. Several procedures can be performed in parallel while others cannot. This issue will be addressed later.

A further issue not pursued in this prototype is the ability to derive the startup procedures from the P&IDs and some knowledge of the chemicals being processed. Several experienced operators said they had performed this derivation for new units they were involved with. The startup procedures for a unit are usually written by the engineers, and the operators run the unit up. They formulate these procedures based on their operational experience. Codifying the knowledge would be of benefit during the design stage of a unit. After the design of a unit, the startup sequence could be determined to identify areas of operational risk.

7.8 Summary

An intelligent system for implementing process startup automation has been successfully developed in an industrial setting. Several test runs have been done and it has performed effectively.

Automating written manuals has improved information retrieval efficiency. Hypertext is useful for management and retrieval of large amounts of information. The success of hypertext in IPSAS has resulted in a new standard for the company. All written manuals that are entered into computers are made available to users through hypertext. The chemical/mechanical state model provides a knowledge representation scheme that is both efficient and understandable. Future development of IPSAS within the company will incorporate this model. The retired expert's startup knowledge was captured and automated within IPSAS. Intelligent system technology has proven to be valuable by capturing expertise knowledge for public use within the company.

IPSAS represents the successful completion of the early steps in the development schedule. The success of these phases of development has resulted in additional funding for a multi-man, multi-year extension of this project at the plant. Additionally, an industrial grant has been awarded to the University of Alberta to support further research.

Automatic startup/shutdown operation has been suggested as a new research direction for process system engineering research. Future work will focus on theory establishment and system implementation as well as industrial applications.

8

Intelligent operation support system for chemical pulp process

8.1 Introduction

This chapter describes the construction of an intelligent operation support system (IOSS) for a batch chemical pulping process. Such a system contains a qualitative reasoning system and a quantitative computing system, and is used to help human operators to produce good quality pulp. The IOSS has many advanced features, such as its *common sense reasoning system* to prevent the system from processing impossible physical data; *variable rules* to organize the knowledge base and reduce both memory space and running time; integrated distributed intelligent system framework to couple both quantitative reasoning and quantitative computation; and a multimedia interface. IOSS can provide the estimated *cooking time* and *kappa numbers*, as well as other process information. Its use has greatly increased interaction between human operators and the controlled process, and thus provides more information interpretation for process operation. IOSS can also be used to train new engineers and operators. This methodology paves the way for process operators to upgrade and improve the old industrial production environment using advanced manufacturing technology.

For a long time, computers have been used for control and monitoring in process industries. With the advent of microcomputer technology, process control is being developed rapidly. Applications of more powerful computers have allowed process engineers to implement more advanced control concepts. The systematic knowledge of process operation has created an environment facilitating the introduction of expert systems (Astrom *et al.*, 1986; Rao, 1991). Existing hardware and programming technologies have matured.

On the other hand, the increasing demands for more effective data processing methods and control strategies to be used in a wider range of industrial applications cannot be met by current computing techniques.

In the past few decades, two important phenomena in scientific discovery and research activities have appeared:

- The most significant research results are often generated in the interdisciplinary field.

- These new scientific discoveries and research results are not replacements of existing technology, just additional modifications or extensions to the current methodology and technology.

As we know, the main objective of most chemical processes is to convert raw materials into high-quality products at the lowest cost, and with low impact on the environment. Therefore, a chemical plant that is well designed, well controlled, and well operated, should meet the above requirements.

The safe operation of a chemical process depends on the human operator and the real-time process control system, and both of them must perform well together. If the processes can be well described by mathematical models, then they can be controlled very well numerically by real-time control systems, thus few human skills and expertise are usually required for the desired operations. However, many chemical industrial processes are very complex in nature, and the development of mathematical models for process control is so difficult that the performance of real-time control systems is poor. In such a case, only the most experienced and skilled operators can handle the operation.

In many pulp and paper mills, high-quality pulp is often produced by a chemical process under batch operation. This chemical process utilizes *cooking liquor* (*cooking acid*) to dissolve and remove the lignin from *wood chips* for the production of chemical pulp. Its primary objective is to cook wood chips to a desirable degree of delignification (i.e. the desired *kappa number*). When a batch digester is used, the quality of pulp is only sampled and tested at the end of the batch cycle.

As a batch digester operates at high temperature and high pressure, it is not practical to remove a pulp sample and analyze its quality (kappa number) on-line. For these processes, optimum operation and quality control depend heavily on the operators who need to use their experience and skills to estimate pulp quality and decide the time to terminate the cooking process in order to produce good quality pulp.

Since today's computers are extremely powerful and can be obtained at affordable prices, artificial intelligence (AI) techniques are being used more and more by researchers and engineers. It is suggested that AI techniques could be used for the operation and control of chemical processes at the human operation level, thus providing a programming methodology for solving ill-structured problems that are too difficult to handle by purely numerical algorithm methods. It is one way to assist human operators in evaluating the data and to advise them how to react to changing conditions. Since expert systems incorporate the knowledge of experienced operators and designers, the operators have the benefit of expert advice on short notice at all times.

In this chapter, an intelligent operation support system (IOSS) for a chemical batch pulping process is introduced. The contents are organized in the following manner. We first present some background information about chemical pulping at a real industrial plant to introduce the application of an intelligent operation support system (IOSS). Then we discuss the problem definition, project strategy, knowledge acquisition and representation. The system configuration based on the concept of integrated distributed intelligent system and the construction of the IOSS are discussed. Finally, evaluations and illustrations are presented, and conclusions drawn.

8.2 Chemical pulping process operation

The industrial process investigated in this project is one of Fraser's sulfite mills which is located in Edmundston, New Brunswick, Canada. This pulping process is fairly complicated. It operates on eight batch digesters and produces approximately 500 tons of unbleached chemical pulp everyday. Digester #4 is the unit used in our case study.

A digester operation flow diagram is shown in Figure 8.1. There are nine major operations in a cooking cycle, starting with loading wood chips into the digester and ending at the dumping stage when the cooking vessel is emptied.

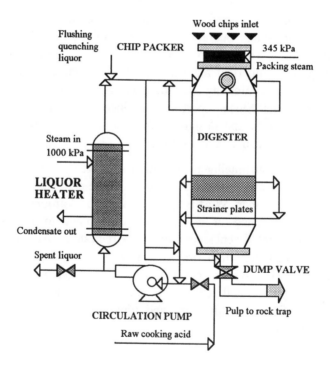

Figure 8.1 Digester flow diagram.

A plot of temperatures and pressure versus time for a cooking cycle is shown in Figure 8.2. There are two temperature curves (one for the region above the strainer plates, and the other for the region below the strainer plates), and one pressure curve. As these curves are plotted on a real-time mechanical recorder strip chart, a physical lag between the pens is required to prevent inter-pen obstruction. This lag represents a 10 minute interval between each plotting pen. This figure reflects all the required operations for a complete batch cooking cycle. These required operational procedures are described in the following paragraphs.

Operation 1: *chip filling*

Wood chips from a wood room and/or from a chip pile are transported to the chip bins located above the digester house, then sent to the digester through a chute. During chip filling, steam at 345kPa is injected tangentially near the top of the digester; this creates a rotating motion of the chips to produce a dense packing per batch. The average duration for the *steam packing* process is approximately 25 minutes.

Operation 2: *capping*

Once the digester is filled with wood chips, the chute is removed. A gasket is placed over the digester opening, the top cover is replaced and secured with bolts. Under normal conditions, the moisture content of the wood chips loaded into the digester is approximately 51%.

Operation 3: *acid filling*

After the digester has been completely sealed, the vessel is ready for the acid filling phase of operation. The operator starts the cooking process, and cooking acid enters the digester at a rate of approximately 1250m^3 per hour. Meanwhile, a cooking liquor sample is taken, and its quality (total %SO$_2$, free %SO$_2$, combined %SO$_2$, and pH) is analyzed. One digester can be filled only once. The acid enters the strainer plates, thus flushing these plates. After a certain amount of acid has entered the vessel, the circulation pump starts automatically.

Operation 4: *indirect steaming*

A digester is equipped with a steam-fed heat exchanger where the circulating cooking liquor is heated. Its temperature is controlled by manipulating the steam going to the liquor heater.

Operation 5: *pressure impregnation*

The pressure impregnation operation forces the cooking acid to penetrate the wood chips. During the rising time (while increasing the temperature), an impregnation pump and other equipment are used to achieve the desired cycling pressure as shown in Figure 8.2. This operation widens the membrane pores in the wood chips, and thus facilitates the penetration of the cooking acid into the chips.

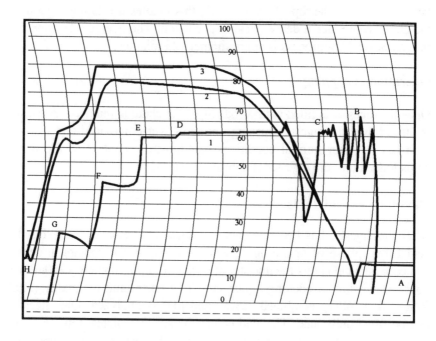

Figure 8.2 Temperature and pressure curves of a cooking cycle.

Operation 6: *side relief*

The side relief operation makes the ratio of cooking acid strength (total SO_2) to air dry weight of wood equal to 18% by removing a calculated amount of cooking liquor from the digester. It also forms a vapor space inside the vessel to safely maintain the high pressure that is required for delignification. Side relief is programmed to start when the temperature of the digester reaches 110°C. The side relief liquor is returned to the accumulator by the circulating pump.

Operation 7: *drain down*

After the operator estimates that the desired pulp quality has been reached, the cooking liquor is drained from the digester. This operation is called *drain down*. It not only reduces the total pressure in the vessel, but also makes room for the quenching liquor, which is introduced in the next phase of operation. A well judged drain down operation is most important because it heavily affects pulp quality.

Operation 8: *quenching*

During the quenching phase, a cold weak liquor is introduced into the digester to further reduce pressure and temperature in the vessel. This weak liquor has practically no chemical strength and comes from the washing and screening processes of the plant. The quenching operation stops the cooking process to prevent dissolving of the cellulose portion of the wood chips, thus preventing loss of yield, and loss in pulp brightness and strength.

Operation 9: *flushing and dumping*

After quenching operation, the *dump valve* is opened and the vessel is emptied. The pulp passes through several processes for further refinement including *rock trap, defiberizer, washing and screening, bleachery,* and *paper machines* (Figure 8.3). If the digester is not emptied immediately after the quenching phase, pulp brightness, pulp yield, and pulp quality would be affected. During the dumping operation, the consistency of the pulp is estimated at 4% fibers, and the stock temperature is under 95°C.

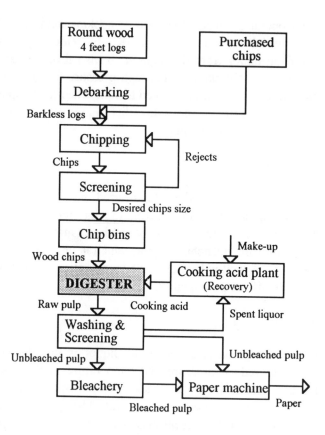

Figure 8.3 Material flow diagram of pulp and paper process operation.

8.3 Problem definition

Digesters cook wood chips to a specified degree of delignification, which is measured by the kappa number. At Fraser, the kappa number is determined by a laboratory test involving titration performed on the pulp sample, since it cannot be measured on-line in the mill. The kappa number can be determined only after the process has been completed through a laboratory test. The lignin content in the wood (pulp) cannot be measured during the batch cooking cycle.

The pulp sample is taken from the dump line after the cooking cycle is completed. The kappa number test is then performed on the sample, and the results are recorded on the digester log sheet. These values are used as a guide by the operators at the bleachery plant for the bleaching operation.

The target kappa number at the Fraser's sulfite mill is 28. A value between 26 and 30 is considered satisfactory. In the case that the laboratory test shows a kappa number higher than 30, indicating a high lignin content still remaining in the pulp, the bleachery operator needs to increase the rate of chlorine/chlorine dioxide for pulp bleaching. A low kappa number for the collected sample (i.e. below 26) indicates loss of pulp strength, pulp brightness, and pulp yield in that particular batch cook. Occasionally, the quality is so bad that the pulp cannot be used for making good quality paper.

At the Fraser's sulfite mill, under normal conditions, a cooking time between 3:45 and 3:55 hours is required to produce a pulp of satisfactory quality (i.e. a kappa number between 26 and 30). Another difficulty is that the qualities of the raw materials (cooking acid and wood chips) that are charged into the digester are not always as anticipated. The cooking acid is produced in an upstream *recovery* process, and its quality changes significantly, which affects the cooking time (i.e. reaction time). The chip quality, which varies in terms of moisture content, species, and chip age, has a similar effect on cooking time. For example, a certain quality of raw material will require a cooking time of 3:35 hours for the digester to achieve its objective, whereas under different circumstance, the batch cooking cycle might require four hours to obtain satisfactory pulp quality.

The operators' objective is to control the digester to produce the best pulp based on the given raw materials by choosing the appropriate operating conditions. The operator must decide when to terminate the process by pressing the drain down button so that good quality pulp can be obtained in the digester at the end of the process. Since no general mathematical model is available for this process, successful operation relies mainly on the operator's private knowledge (such as personal experience, expertise and heuristics). However, such experienced operators are few. The frequent shift operation among the eight digesters very often confuses operators, even experts. Table 8.1 shows a statistical analysis of the performance of the batch cooking operation in July 1990 at Fraser.

Table 8.1 Performance analysis of batch cooking operation (July 1990)

Quality	Cook number	Kappa number	Percentage
Best	114	28	12.7 %
Satisfactory	334	26~27	46.1 %
		29~30	
Poor	< 124	> 30	21.1 %
	> 153	< 26	17.1%

Table 8.1 shows that out of 725 batch cooks during that month, only 114 cooks (15.7%) had reached the target kappa number (i.e. 28). There were 277 (38.2%) poor quality cooks, which indicates that the digester operation and quality control are not satisfactory.

There are several factors that cause 38.2% of the cooks to be poor quality. A main reason is wrong interpretation of the process data by a human operator who is responsible for all eight digesters during a shift of 12 hours. Even operators who have many years of operations experience and know the process very well, are inefficient at quickly recalling the effects of particular data on one digester and different data on another digester. Thus, they are confused in operation.

Again, the main criteria to be satisfied to produce pulp of the desired quality is to choose the right cooking time and to maintain optimal operating conditions. Cooking time is a function of the raw materials quality (wood chips and cooking liquor) and the operating conditions (temperature and pressure).

Until recently, we have found no successful means of capturing the expertise from those very experienced operators who have spent many years at the plant, and accumulated much private knowledge. This very valuable knowledge could be lost with the retirement of these experts. How to codify their experience and expertise for future public use, i.e. to store these most valuable assets for the company, has become an important and challenging task. In summary, there are the following difficulties in the operation and control of the current chemical batch pulping process:

- There are many ill-structured problems that are difficult to solve by mathematical modeling plus numerical control methods.

- In some mills, the kappa number can be estimated only by a few experienced operators, which greatly affects the pulp product quality.

- There exists little interaction between the process operators and the controlled process.

- The digesters are not always at the optimal operating conditions.

- Valuable private knowledge about the pulping operation cannot be accumulated, then transferred for public use.

8.4 Problem-solving strategy

In order to solve the problems encountered above, our strategy is to develop an intelligent operation support system (IOSS) that suggests the proper cooking time based on the available information concerning the process, and monitors the cooking process to estimate kappa numbers. By investigating the effect of all the available process quality inputs on the cooking time (private knowledge), the IOSS can quickly process the knowledge and provide operational support. The private knowledge is obtained from the experts (operators), and coded into the computer system. A mathematical cooking model developed based on S_m-factor (public knowledge) can also be implemented in such a computer program, and used for supervisory control.

This methodology also aims at increasing the interaction between the process and human operators, providing more information interpretation during the pulping operation, thus optimizing the production environment. It transfers and accumulates the expert knowledge into computer programs so that those who do not have the operational experience can also control digesters at the level of expert operators. Obviously, using the newest developments in advanced technology to upgrade and improve the old industrial manufacturing facility is another advantage.

8.5 Knowledge acquisition

It is important for the developers to understand in detail the digester's operation and the related problems before starting the construction of an IOSS. The main task for the developers is to acquire the valuable knowledge that has been used in the digester's operation to produce satisfactory products. Then, the acquired knowledge can be built into the IOSS so that the knowledge can be easily transferred to the users.

Another task of the IOSS is to increase the interaction between operators and the controlled process and give operators assistance to produce good quality pulp. The more accurate and the better organized the knowledge is, the more efficient the IOSS is.

Most of the knowledge we obtained is classified as private knowledge that is very specific to the process behavior and cannot be obtained from textbooks or literature. This knowledge, which is mainly acquired from the operation experts (four operators and one superintendent were interviewed), is the most important and valuable information.

During each interview, it was important to explain this project to experts. Some operators were concerned about their job security. Technically, the IOSS is a computer program that solves problems at the experts' level. It is important to tell operators that human being are much better than computers at the reasoning level (Surko, 1989), and the objective of the IOSS is to support them in achieving much better operation and production of good quality pulp, rather than letting operators sell out their knowledge and experience to the management, then lose their jobs.

8.6 Qualitative knowledge representation

Once the process operation and related problems were studied, all the available quality inputs and outputs were investigated (Smith and Hubert, 1983). These process variables can be classified as follows:

Process quality inputs:

Wood chips	Cooking acid	Operation conditions
Chip quality	Total % SO_2	Temperature
Chip size	Free % SO_2	Combined pressure
Chip load	% SO_2	Cooking time
	pH	

Process quality outputs:

Pulp quality
Kappa number
Pulp brightness

Table 8.2 shows the desired conditions for all process variables. If a process input is outside these conditions, the desired pulp quality would not be achieved. The impact of the deviation of each process input on the pulping process behavior is described in the remainder of this section.

8.6.1 Chip load

Generally, a batch digester is filled with wood chips up to a specific level. Weight cells installed at the bottom of the digester are used to measure the load of chips. The weight cells are calibrated to provide values between 0% and 100%. A value between 35% and 40% gauge indicates an optimum chip load. A value higher or lower than the desired value indicates that the moisture content is higher or lower, respectively. If low moisture content wood chips are fed into the vessel, less cooking time would be required for the batch cycle because it is easier for the acid to penetrate the chips and to attack the lignin chemically. Under low chip moisture conditions, less steam is required to heat up the system, and a higher brightness of the final pulp usually results.

Table 8.2 Desired conditions for pulping quality variables

Quality inputs	Desired conditions
Load of wood chips	35%~40% gauge
Size of wood chips	1"~3"×1"~3"×1/4"~1/2"
Quality of wood chips	Good quality (< 6 weeks old) 90% softwood~10% hardwood
Total % SO_2 of initial cooking acid	5.45%~6.15%
Combined % SO_2 of initial cooking acid	2.65%~3.00%
Free % SO_2 of initial cooking acid	2.80%~3.15%
pH of initial cooking acid	3.4~3.6
Operating temperature	162~165°C
Operating pressure	675~685kPa
Cooking time of batch cycle	3:45~3:55 (hour)
Quality outputs	**Desired conditions**
Kappa number	26~30
Pulp brightness	62~68

In the event of high moisture content, more cooking time is required. First of all, the acid concentration is reduced, secondly, it takes more time for the acid to penetrate the chips. A higher steam rate is also required with high moisture content chips.

The input identifies the wood chips as *old* chips or the type of wood blend (hardwood or softwood). Old chips are generally easy to digest but produce dark pulp (low pulp brightness). When the chips loaded into a digester are considered old, less cooking time is required as well as a lower steam demand.

At Fraser, the blend of wood chips under the desired condition is approximately 10% hardwood and 90% softwood. If more than 10% hardwood is used in the digester, the cooking time and pulp strength are affected. A high hardwood content (> 10%) in batch cooking results in a shorter cooking time. Higher pulp brightness and lower pulp strength are also obtained as final results. Because hardwood is made of short fibers, if the chip blend contains more than 10% hardwood, problems would eventually occur at the washing and screening processes.

If the wood chips fed into a digester are both old chips and more than 10% hardwood, a much reduced cooking time is required. The pulp brightness is almost unpredictable in this situation, and the strength of the pulp is lower. On the other hand, the steam required to heat-up the system is less.

8.6.2 Chip size

The standard size of a single wood chip is approximately 1"-3"×1"-3"×1/4"-1/2" in dimension. If the average dimension of chip is very different from the standard size, a different cooking time is required for the batch cycle to achieve the right quality pulp.

In the case that the size of chips loaded into a digester is larger than standard, which usually occurs when the chip screening system is malfunctioning, longer cooking time is required because it takes more time for the acid to penetrate the whole chips for delignification. If the chips are too small (i.e. under standard size), less cooking time is required and a lower pulp brightness is obtained in the final pulp product.

8.6.3 Total SO$_2$

The cooking acid that is fed into the digester consists of the following major components:

$$Mg(HSO_3) + SO_2 + H_2O$$

The total SO$_2$ consists of both the combined SO$_2$ and free SO$_2$. The concentrations of these components are measured in percentage sulfur dioxide. The higher the total SO$_2$ percentage of the cooking acid is, the less cooking time is required. The final pulp brightness can also be lower depending on the ratio of the free SO$_2$ and combined SO$_2$. If the total SO$_2$ percentage is low, more cooking time is required to obtain a desired pulp quality.

8.6.4 Free SO$_2$ and combined SO$_2$

The desired condition for the free SO$_2$ and combined SO$_2$ of the cooking acid is that the difference between the free SO$_2$ percentage and combined SO$_2$ percentage is 0.15% to 0.20%. When the free SO$_2$ minus the combined SO$_2$ is less than the desired condition, which indicates low acidity (high pH), more cooking time is required. If the free SO$_2$ minus the combined SO$_2$ is higher than the desired conditions, which causes high acidity (low pH), less cooking time is required.

8.6.5 pH

For the pH of the cooking acid, a low value indicates a higher liquor acidity resulting in a reduced cooking time requirement. Under high cooking acid acidity, the final pulp brightness produced is usually lower. If the pH of the cooking acid is high, indicating low acidity, more cooking time is required, but the pulp brightness is not affected.

8.6.6 Operating pressure

The operating pressure of the cooking cycle is the total pressure inside the digester during the digesting process at operating temperature. As the vessel is completely sealed, the pressure is increased by raising the temperature. The operating pressure is set up at 680kPa and maintained by a pressure relief valve. In some situations, when the pressure relief valve does not work properly or when the pressure set-point is incorrectly adjusted, a different operating pressure is obtained. If the operating pressure is lower than the desired condition, longer cooking time is required. In the situation when a higher operating pressure is used, less cooking time is required and a lower pulp brightness is obtained in the final pulp product.

8.6.7 Operating temperature

The effect of operating temperature is the same as the effect of operating pressure. A low operating temperature results in a longer cooking time, while a high operating temperature results in a shorter cooking time and produces low pulp brightness.

An operating temperature outside desired conditions is mainly due to bad steam quality and misbehavior of the steam control valve.

Table 8.3 shows the required cooking time for different process quality input conditions. All information was acquired from the operators. Because normal cooking time differs by less than 10 minutes (i.e. between 3:45 hours and 3:55 hours), a minimum of 10 minutes range for counting the cooking time is used in order to classify the different input conditions. When more than one quality input are outside the desired condition, these ranges can be compiled together to derive a coupled cooking time (i.e. a near optimum cooking time range).

Table 8.3 Description of each situation

Individual inputs	Description	Average cooking time
Wood chips load	22%~29% gauge	3:20~3:30 hour
	30%~34% gauge	3:30~3:40 hour
	35%~40% gauge	3:45~3:55 hour
	41%~46% gauge	4:00~4:10 hour
	47%~55% gauge	4:10~4:20 hour
Wood chips quality	Good quality	3:45~3:55 hour
	Old chips	3:25~3:35 hour
	more than 10% hard wood	3:25~3:35 hour
	Old chips & more than 10% hardwood	3:15~3:35 hour
Wood chips size	Standard size	3:45~3:55 hour
	Over standard size	4:05~4:15 hour
	Under standard size	3:25~3:55 hour
Total %SO_2 of initial cooking liquor	4.00~4.80%	4:05~4:15 hour
	4.81~5.44%	3:55~4:05 hour
	5.45~6.15%	3:45~3:55 hour
	6.16~6.50%	3:35~3:45 hour
	6.51~7.40%	3:25~3:35 hour
pH of initial cooking liquor	2.5~2.9	3:25~3:35 hour
	3.0~3.3	3:35~3:45 hour
	3.4~3.6	3:45~3:55 hour
	3.7~4.0	3:55~4:05 hour
	4.1~4.6	4:05~4:15 hour
Operating temperature	154~157 °C	4:10~4:20 hour
	158~161 °C	4:00~4:10 hour
	162~165 °C	3:45~3:55 hour
	166~169 °C	3:30~3:40 hour
	170~175 °C	3:20~3.30 hour
Operating pressure	575~624 kPa	4:05~4:15 hour
	625~674 kPa	3:55~4:05 hour
	675~685 kPa	3:45~3:55 hour
	686~735 kPa	3:35~3:45 hour
	736~785 kPa	3:25~3:35 hour

8.7 Quantitative computation

In the operation and control of the cooking process, both qualitative and quantitative analyses are needed. Usually, qualitative decisions are efficiently based on symbolic and graphic information, while quantitative analysis is more conveniently performed using numerical information. Each method often complements the other (Rao, 1991).

The IOSS has a quantitative computation system to calculate cooking time and kappa number. The mathematical model implemented in the IOSS is based on the $S_{\text{m-factor}}$. This system is programmed in the FORTRAN™ language and its module is called 'cookmodel.for'. The results from 'cookmodel.for' are precise and superior to those from a qualitative reasoning system. However, this numerical computing system can be used only when the chip size and chip quality are under normal conditions. The qualitative reasoning system using heuristics and private knowledge cannot guarantee perfect solutions. Therefore, it solves ill-formulated problems and generates suggestions about possible changes in the operating environment.

The following sections describe the computation process to calculate the $S_{\text{m-factor}}$, and the equations to calculate the cooking time and kappa number.

8.7.1 Models for kappa number estimation

In the sulfite pulping process, the S_{factor} method (Bylund and Thorsell, 1982) is often used to calculate the degree of delignfication. The S_{factor} model includes both temperature and pressure. Its derivation procedure is given as follows.

The reaction rate of lignin dissolution in sulfite pulping can be written as follows (Bylund and Thorsell, 1982):

$$r_L = \frac{d[L]}{dt} = -k(t) \cdot [HSO_3^-]^\sigma \cdot [H^+]^\beta \cdot [L]^\gamma, \tag{8.1}$$

where

r_L = rate of lignin dissolution,

$[L]$ = lignin content in wood residue,

$k(t)$ = temperature-dependent rate constant for lignin dissolution,

$[HSO_3^-]$ = hydrogen sulfite ion-concentration of cooking liquor,

$[H^+]$ = hydrogen ion-concentration of cooking liquor, and

σ, β, γ = parameters.

By making the following assumptions, equation (8.1) can be simplified.

(a) The reaction orders with respect to $[H^+]$ and $[HSO_3^-]$ are of the same magnitude (i.e. $\sigma = \beta$).

(b) The product $[H^+] \cdot [HSO_3^-]$ is directly proportional to the partial pressure of SO_2 gas in the digester.

According to Bylund and Thorsell (1982), both assumptions are approximately valid. After rearrangement of equation (8.1), the following equation is obtained:

$$\frac{d[L]}{[L]^g} = -k(t) \cdot P_{SO_2}^\sigma \cdot dt. \tag{8.2}$$

The right-hand side of equation (8.2) includes three main cooking parameters:

t = temperature of digester contents in which an Arrhenius expression goes into reaction rate constant, as the H_{factor} for kraft pulping, and

P_{SO_2} = partial pressure for SO_2 inside the digester.

By making one more assumption, equation (8.2) can be easily integrated.

(c) The reaction order with respect to lignin concentration in the wood residue is of pure first order (i.e. $\gamma = 1.0$). Then, a new equation is obtained:

$$\ln[L]_t - \ln[L]_0 = -\int_0^t k(t) \cdot P_{SO_2}^\sigma \, dt . \tag{8.3}$$

The S_{factor} is defined by the following equation:

$$S = \int_0^t k(t) \cdot P_{SO_2}^\sigma \, dt , \tag{8.4}$$

and

$$k = e^{[A-B/T]}, \tag{8.5}$$

where

$S = S_{factor}$,
A = parameter ($B/373$), and
B = activation energy parameter (function of wood species).

Based on laboratory experimental study, the following numerical values are obtained (Yorston and Liebergott, 1965):

$\sigma = 0.75$ to 0.85 bar, and

$\beta = 10166$ Kelvin.

Using experimental trials from a sulfite mill at Domsjo in Sweden, Bylund and Thorsell (1982) obtained the following values:

$\sigma = 0.75$ bar, and

$\beta = 11500$ Kelvin.

Empirical models for the prediction of the kappa number have been built using the S_{factor} for sulfite pulping (Bylund and Thorsell, 1982; Haywood, 1989). These models are as follows:

$$\text{kappa number} = a_1 + a_2 \cdot C + a_3 \cdot e^{-a_0 \cdot S} + a_4 \cdot C \cdot e^{-a_0 \cdot S}, \tag{8.6}$$

where

$S = S_{factor}$,

C = concentration of cooking liquor, and

a_0, a_1, a_2, a_3, a_4 = regression coefficients.

And

$$\text{kappa number} = a_1 - a_2 \cdot P_{tot} \cdot \int_0^t k \cdot dt + a_3 \cdot C + a_4 \cdot D, \qquad (8.7)$$

where

P_{tot} = total pressure in the digester,

C = acid concentration,

D = % total SO_2 of acid, and

a_0, a_1, a_2, a_3, a_4 = regression coefficients.

The S_{factor} is accurate under laboratory conditions where all the cooking factors (cooking liquor quality, wood chip quality, temperature, etc.) can be controlled. In a real plant, the S_{factor} alone is not adequate (Haywood, 1989). This can be confirmed from the fact that Bylund and Thorsell obtained a regression coefficient R = 0.96 using laboratory data for equation (8.6) (1982) and a regression coefficient R = 0.84 using real plant data for equation (8.7) (Haywood, 1989).

It should be noted that all these mathematical models do not have wood chip quality as a dependent variable. It is well established that moisture, mixed wood species, contamination, and chip dimension among other variables have a strong impact on pulp quality (Bylund and Thorsell, 1982; Andrews and Barnes, 1985; Parker, 1988). If these input variables cannot be monitored or are unknown, the results from these mathematical models will be inaccurate.

There is no developed mathematical model that takes into account chip size, chip quality, chip moisture, and so on. As these input variables can affect the cooking process significantly, a mathematical model cannot always be employed to estimate kappa numbers and cooking time,

$$\text{kappa number} = a_1 + a_2 \cdot \int_0^t P_{\text{tot}} \cdot k \cdot dt + a_3 \cdot [TS(0)] + a_4 \cdot [TS(10)] \quad (8.8)$$

where

a_0, a_1, a_2, a_3, a_4 = regression coefficients,

k = relative reaction rate,

P_{tot} = total pressure inside the digester (atm),

$TS(0)$ = percentage of total SO_2 of the cooking liquor during acid filling (before cooking),

$TS(10)$ = percentage of total SO_2 of the circulating cooking liquor 10 minutes after the operating temperature has reached a maximum.

The modified S_{factor} is represented by the following equation

$$S_{m\text{-}factor} = \int_0^t P_{\text{tot}} \cdot k \cdot dt . \qquad (8.9)$$

The relative reaction rate (k), can be calculated using the following equation

$$k = e^{[F - \frac{E}{R \cdot T}]}, \qquad (8.10)$$

where

E = activation energy (cal/mol),
F = frequency factor (constant),
R = gas constant (cal/mol-K), and
T = absolute cooking temperature (K).

For sulfite pulping, the above parameters can be chosen to have the following numerical values:

E = 20200 cal/mol,
F = 27.25, and
R = 1.987 cal/mol-K.

The numerical values of the regression coefficients in equation (8.8) are given as follows:

$a_1 = 108.2,$
$a_2 = 0.02170,$
$a_3 = 8.336,$ and
$a_4 = 7.425.$

The regression analysis based on the above parameters results in a coefficient of multiple correlation (R) of 0.958.

Now, by setting the kappa number to 28 (i.e. target kappa number at Fraser's mill), equation (8.8) can be rearranged in the following manner to calculate the $S_{\text{m-factor}}$:

$$S_{\text{m-factor}} = \frac{28 - a_1 - a_3 \cdot [TS(0)] - a_4 \cdot [TS(10)]}{a_2} \tag{8.11}$$

The value for '$TS(0)$' can be obtained during acid filling and the value for '$TS(10)$' can be obtained from the circulating liquor during the cooking cycle (e.g. approximately 2:10 hours after the cooking cycle has started). Once both values are obtained, the target $S_{\text{m-factor}}$ can be calculated using equation (8.11).

During the rising time operation stage, when the temperature and total pressure are increased to their operation conditions, the $S_{\text{m-factor}}$ is calculated using equation (8.9), and is represented by $S_{\text{m1-factor}}$. However, when the temperature and total pressure are at their operating conditions (they do not change with time), the $S_{\text{m2-factor}}$ can be calculated by a simplified equation as:

$$S_{\text{m2-factor}} = P_{\text{tot}} \cdot k \cdot \Delta t. \tag{8.12}$$

By using equation (8.9) for the 'rising time' operation of the process and using equation (8.12) for the remaining part of the process, the total $S_{\text{m-factor}}$ can be calculated:

$$S_{\text{m-factor}} = S_{\text{m1-factor}} + S_{\text{m2-factor}}. \tag{8.13}$$

Equations (8.8) to (8.13) are used in the IOSS to predict the cooking time, kappa number, as well as $S_{\text{m-factor}}$.

8.7.2 Computation process

The computation process of the quantitative system is shown in Figure 8.4. The required data from the *database* array and the *s_fact* array (both are in a database) are passed to the numerical computation module by the inference engine to calculate the target $S_{\text{m-factor}}$. Here, the target $S_{\text{m-factor}}$ is calculated based on the target kappa number (i.e. 28). Total $SO_2(10)$ is required for the computation of $S_{\text{m-factor}}$, which is obtained 10 minutes after the operating temperature has reached the maximum value. Usually, it takes approximately 2 hours for the temperature to reach a maximum. In other words, the target $S_{\text{m-factor}}$ can only be calculated approximately 2:10 hours after the cooking cycle has been started.

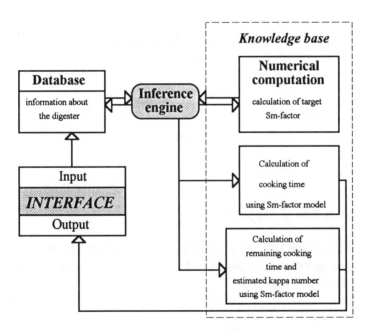

Figure 8.4 Computation of quantitative system.

Once a target $S_{m\text{-factor}}$, is calculated, the inference engine passes the calculated $S_{m\text{-factor}}$, to different modules to calculate cooking kappa number. The result is then displayed on the computer screen through the user interface.

8.7.3 Calculation of cooking time

To calculate the cooking time, the system requires the $S_{m1\text{-factor}}$ and the time at which the cooking liquor sample test for the total SO_2 (10) is taken. $S_{m1\text{-factor}}$ is computed at different time intervals. Once the temperature and pressure of the digester are held constant at their operating conditions, the $S_{m2\text{-factor}}$ can be calculated using equation (8.9). Therefore, from equations (8.12) and (8.13), the required cooking time for the production of good quality pulp can be calculated, i.e.

$$\text{Cooking time} = \left[\frac{S_{m\text{-factor}}(\text{target}) - S_{m1\text{-factor}}}{P_{tot} \cdot k} \right] + t_1, \qquad (8.14)$$

where

t_1 = the time of which the total $SO_2(10)$ is taken (hour),

P_{tot} = operating pressure (atm), and

k = relative reaction rate.

8.7.4 Estimation of kappa number

When the process is at its operating temperature and pressure, the user can obtain an estimated current pulp kappa number. The kappa number can be calculated using equation (8.8) based on the value of the $S_{mt\text{-factor}}$ at the requested time. $S_{mt\text{-factor}}$ can be calculated using the following equation:

$$S_{mt\text{-factor}} = S_{m1\text{-factor}} + P_{tot} \cdot k \cdot [t - t_1], \qquad (8.15)$$

where t = any particular time after the target $S_{m\text{-factor}}$ is calculated.

The remaining cooking time can also be calculated from the following equation:

remaining cooking time = cooking time - t.   (8.16)

All results produced by the quantitative computation system can be used by the operators for on-line supervisory control.

8.8. System construction

In the IOSS, both qualitative and quantitative knowledge are integrated and coordinated by a meta-system (Rao, 1991). Symbolic reasoning, numerical computation as well as graphical representation are integrated to facilitate the functions of the IOSS. Figure 8.5 demonstrates this configuration. Such architecture facilitates access of these two types of knowledge to an end user.

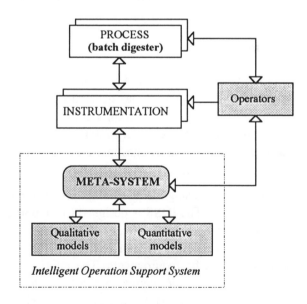

Figure 8.5 Integrated architecture in IOSS.

Both qualitative and quantitative systems can be used by the operators for decision-making. However, as the mathematical model is only a function of the pressure, temperature, total SO_2, and time, the quantitative computation system cannot be used when other input variables (i.e. chip load, chip size, chip quality, and pH of acid) are outside their normal quality conditions. Also, the quantitative computation system does not provide qualitative information about the effect of input variables on the final pulp product (such as effect on pulp brightness, pulp strength, etc.).

As the numerical computation generates rigid solutions, the quantitative computation system should be used first to predict the required cooking time for the production of good quality pulp, if the mathematical model is applicable.

The qualitative reasoning system can be used by the users to obtain quality information about the process behavior. The qualitative system does not guarantee perfect solutions, but it can give the directions concerning the process behavior changes. Such a system increases the interaction between operators and the actual process (digester).

8.8.1 Implementation platform

As mentioned before, the majority of the knowledge base of the IOSS is coded in C™ programming language, and developed using the Borland TURBO C™ compiler. The numerical computation program is written using FORTRAN™ language. There are advantages and disadvantages to using conventional programming languages instead of using a commercial expert system development shell.

The primary advantage of using a conventional programming language is that it provides flexibility. The IOSS is designed to support the digester operation and to meet the operators' demands. By using a conventional programming language, the developer can customize the program, and modify it for the users' specific requirements. The interface of the IOSS consists of a *main-menu*, many *sub-menus*, and different windows, which make the input/output of information much easier to interchange between the user and the IOSS. A well developed I/O interface and easy-to-use computer program environment are very important for the IOSS to be accepted by end users.

Another advantage of using a conventional programming language is that once the system (IOSS) is completed and compiled, it can be installed on any PC-based computers without the purchase of any software or shell.

The main disadvantage of using conventional programming languages to build intelligent systems is that the development time is usually much longer than when using a commercial expert system shell. Many commercial packages provide built-in interface, and the structure of the knowledge base is already designed. Only the knowledge has to be entered into the system. This saves a lot of development time.

8.8.2 Function modules of IOSS

The IOSS is a menu-driven system, which can be easily installed on any PC/MSDOS platform. The main function modules contain the following choices that can be selected by the user on the computer screen. There are two main menus. They are:

Main-menu #1

[Intro] [Pulping] [speC] [Inputs] [Situation] [Option] [Next menu]

Main-menu #2

[Prev. menu] [Operation] [Flow dia] [Shutdown] [Startup] [Editor]

As shown in Figure 8.6, when the IOSS is running, main-menu #1 appears on the top of the display screen. Each selection from the main menu provides a sub-menu and/or a window system to display information to the user, or to receive information from the user. Some selections are linked together to process the information better. The main functions of each selection are described as follows:

[Intro]: This selection provides a brief introduction about the IOSS. It gives the user a quick understanding of each selection from main-menu #1 and main-menu #2.

[Pulping]: The Pulping selection gives an introduction about chemical pulping. It describes the purpose of digesting and the concept of delignification.

[Spec]: This sub-menu provides specifications for the equipment related to the chemical pulping process. It provides, for example, the type of pump with its capacity, or the size/material of a tank for a selected system. The IOSS provides information for the following equipment and systems: digesters, acid system, condensation system, water system, drain down system, dump system, and flushing liquor system.

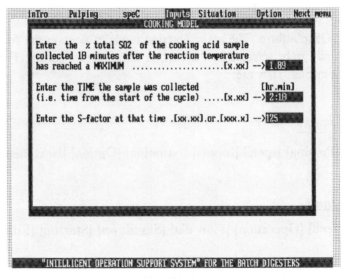

Figure 8.6 Input data for quantitative computation systems.

[Inputs]: The Inputs module takes the data entered by the user into the IOSS. The data are held in a database. These data are the process quality inputs that describe a batch cooking cycle. These inputs are presented in

Table 8.2. The data required by the $S_{\text{m-factor}}$ model are also entered from this sub-menu.

[Situation]: The Situation module provides qualitative knowledge as well as quantitative knowledge based on the status of the database (i.e. information entered through the Inputs selection). The qualitative knowledge includes the estimated cooking time and remaining cooking time, the effect of the quality inputs on the quality outputs, and the possible reasons for which a certain quality input is outside its desired operation conditions. The quantitative knowledge is the results from the $S_{\text{m-factor}}$ model that provides the on-line computed cooking time, remaining cooking time, and estimated kappa numbers during a cooking cycle.

If no information is entered into the database by the user prior to selection of the Situation, the IOSS will utilize the default values attributed to the database.

[Option]: This selection has a similar function as the previous one (i.e. Situation). The Option selection can provide cooking time by using a bar graph. It also recommends the cooking time of coupled quality inputs. This module can also display the desired conditions or the current conditions of all process quality inputs in one single window.

[Next menu]: By choosing the Next menu selection, the top bar menu will be changed from Main-menu #1 to Main-menu #2.

[Prev. menu]: By choosing the Prev. menu selection, the top bar menu will be changed from Main-menu #2 to Main-menu #1.

[Operation]: This module provides all the operations to be performed for a single batch cooking cycle.

[Flow dia]: This selection provides a graphical display of the batch digester system which includes the main piping, circulating pump, dump valve, and liquor heater. It also provides the initial cooking liquor quality, the initial chip quality, and the direction of the circulating liquor.

[Shutdown]: This selection provides the required steps on how to shutdown the operation of digesters.

[Start-up]: This selection provides the required steps on how to start-up the operation of digesters.

[Editor]: This Editor can be accessed the user. The Editor used in the IOSS is PE2™, which is an IBM personal computer DOS™ editor. With PE2™, we can directly edit the files that are required for the proper operation of the IOSS. These files contain information about pulping, process specifications, process operations, and shutdown/startup procedures.

8.8.3 Database

The database of the IOSS is defined as an array containing the data of each process quality input. These values represent the status of a batch cooking process. The database array is given as 'float database[10]'.

As the configuration of the IOSS is separated into several different subsystems, the database is defined as a global array so that more systems can use it. The specified float means that the data in the database are held as the real type data. The following example shows that all the process quality inputs are contained in the database.

Example 1: The setup of the database array

```
database[0] = 38;        /* CHIP LOAD */
database[1] = 1.0;       /* CHIP QUALITY */
database[2] = 1.0;       /* CHIP SIZE */
database[3] = 5.75;      /* TOTAL SO₂(0) */
database[4] = 3.50;      /* pH */
database[5] = 163;       /* TEMPERATURE */
database[6] = 680;       /* PRESSURE */
database[7] = 2.80;      /* COMBINED SO₂ */
database[8] = 2.95;      /* FREE SO₂ */
database[9] = 0.00;      /* TOTAL SO₂(10) */
```

All the values in the above example are default values. These values are automatically entered into the database when the IOSS starts running. However, each individual input can be modified by a user who selects the

Inputs from Main-menu #1. Each individual input can be selected from the sub-menu and its new values can be entered from a keyboard. All values in the database array except 'database [9]' are used to process qualitative knowledge. Four values (temperature, pressure, total $SO_2(0)$, and total $SO_2(10)$) are used to process quantitative knowledge.

There is another array which contains information about process behavior, and is defined as 'float s_fact[4]'.

This array is also a part of the database, and contains other required data for the quantitative computation system (i.e. $S_{m\text{-factor}}$ model). Example 2 shows the information contained in the s_fact array.

Example 2: Setup of s_fact array

s_fact[0]	/* $S_{m\text{-factor}}$ (target) */
s_fact[1]	/* $S_{m\text{-factor}}$ (10 minutes after maximum temperature is reached) */
s_fact[2]	/*Time when a cooking liquor sample for total $SO_2(10)$ was taken */
s_fact[3]	/* Estimated total cooking time */

The $S_{m\text{-factor}}$ (target variable) is calculated by the mathematical model based on both a kappa number of 28 (target quality criterion) and the input data from the database array. The value for s_fact[1] is entered by the user and consists of the $S_{m\text{-factor}}$ at the time when the sample for the total %SO_2 (10) is collected. This $S_{m\text{-factor}}(10)$ can be obtained from another computer system which is currently physically connected to digester #4 as a data acquisition system. Such a system can calculate on-line $S_{m\text{-factors}}$. The s_fact[2] contains the time when the sample for total $SO_2(10)$ is taken. With the use of the first three numerical data, a cooking time can be calculated. The calculated cooking time is derived from the $S_{m\text{-factor}}$ model and held in s_fact[3].

Also from the information contained in the s_fact array, the remaining cooking time and estimated kappa number can be calculated and provided to the operator for supervisory control.

8.9 Knowledge base organization

The qualitative knowledge for each quality input variables is held in different arrays and matrices. Each array or matrix contains a different type of quality information, and is kept separately. The knowledge contained in these arrays and matrices are described below.

float ranges[][2]

Defining a matrix as 'float ranges[][2]' means that its dimension is [x] by [2], where x can be any positive integer. Therefore, 'ranges[][2]' has no specified dimension in the x direction, and it can be very large, up to exceeding the limitation of computer memory.

The data contained in this matrix are the different possible values of each process input variable the IOSS can process. Most process input variables are separated in a maximum of five different quality ranges. The values contained in 'ranges[][0]' are the lower limit of each range, and the values contained in 'ranges[][1]' are the upper limit of each range. Having no specified dimension for the range matrix makes the system flexible and easy to modify. The current system can manage up to ten different values for each process input variable.

char *status[10]

The *status* array contains quality information about the input variables. Each input variable has its own status array. The quality information contained in the status array is a word such as *high*, *low*, or *desired*.

char *time_required[10]

The *time_required* array contains the estimated required cooking time. Each value in the ranges matrix has a corresponding cooking time.

char *situation[10]

All other quality information about different values of each variable are held in a *situation* array. The information contained in this array are, for example, the results of the final pulp quality (e.g. high, low, or good pulp brightness) as well as the possible reasons why the value of the input variable is outside the desired quality range. In some cases, the suggestions (e.g. what to do) are stored in this quality range (situation array).

It should be mentioned that currently, the information contained in the above arrays and matrices are displayed to the user through a window system.

char *time_limits[][2]

As the IOSS provides the user graphical displays (e.g. bar graph) to show the effect of each input variable on the cooking time, an array *time_limit* is used to hold the supper limit and lower limit of an estimated range for cooking time. For example, if 'time_limits [][0] = +20', and 'time_limits [][1] = +10', then the bar on the *bar graph* for that particular input variable will show a yellow colored section indicating that the estimated cooking time is about 10 to 20 minutes more than the normal required cooking time.

float weight[10]

The *weight* array contains weight values for input variables. These weight values are used by the bar graph to couple the effect of several input variables on the cooking time. When a reduced cooking time is required for the cooking cycle to achieve the desired pulp quality, the corresponding weight value in the weight array contains a negative (-ve) value. When a longer cooking time is required, the weight is a positive (+ve) value. Under the desired condition, the weight value is equal to zero.

The knowledge base of the qualitative reasoning system is coded in a separate module called 'KNBASE.C'. If the knowledge needs to be updated or modified, the developer can add, delete, or modify the content of the arrays and matrices. The knowledge base can be easily modified without changing the logic or the rule structure. Even people who do not have a background in C™ language can accomplish the modification of knowledge base in the IOSS.

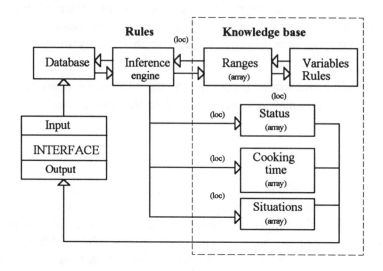

Figure 8.7 Reasoning process of qualitative system.

The reasoning process of the qualitative system is described in Figure 8.7. The process variable data are entered into the database by the user. These data are processed in the knowledge base by the inference engine. The knowledge base, which consists of arrays and matrices, receives the data in the module containing the ranges matrix. The ranges matrix and the values of that particular input are processed through a variable rule system to find the location (loc.) in the ranges matrix that satisfies the condition. The location (loc.) is returned to the inference engine. The inference engine feeds the location (loc.) back to the knowledge base for further processing by the remaining arrays (i.e. status, time_required, and situation). The information in these arrays at the location (loc.), which contains the qualitative knowledge corresponding to the value of the input variable being processed, can be displayed on the computer screen. The information provided by the qualitative reasoning system can be used by the human operators for on-line supervisory control.

8.10 Other advanced system features

8.10.1 Coupling system

If some input variables are not within their desired values, the system needs to couple the effect on the cooking behavior and to derive a coupled required cooking time. Figure 8.8 shows how the input variables can be coupled together. Thus, different cooking times can be obtained at three different levels.

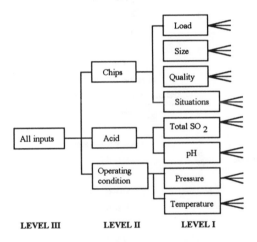

Figure 8.8 Coupling input variables.

Level I derives cooking time for each individual input variable. Level II provides the coupled cooking time for the wood chips (the system couples the effects of chip size, chip load, and chip quality together), the cooking acid (the system couples the effects of 'Total SO_2' and 'pH' together), and the operating condition (the system couples the effects of operating temperature and operating pressure together). Level III provides a coupled cooking time for all input variables together. The coupled cooking time is provided to users thought the bar graph display under the Options selection in Main-menu #1. The compilation of the effect of these input variables on the cooking time is done by adding the weight values in the weight array of each input variable. Table 8.4 gives the weight values for the different possible values of all input variables.

Table 8.4 Description of each situation at level I

Inputs quality	Description of each situation (rules)	Weights
Wood chips load	22~29% gauge	-4
	30~34% gauge	-2
	35~40% gauge	0
	41~46% gauge	+2
	47~55% gauge	+4
Wood chips quality	Good quality	0
	Old chips	-3
	More than 10% hard wood	-3
	Old chips and more than 10% hardwood	-30
Wood chips size	Standard size	0
	Over standard size	+3
	Under standard size	-3
Total SO_2 of initial cooking liquor	4.00~4.80%	+3
	4.81~5.44%	+1
	5.45~6.15%	0
	6.16~6.50%	-1
	6.51~7.40%	-3
pH of initial cooking liquor	2.5~2.9%	+3
	3.0~3.3%	+1
	3.4~3.6%	0
	3.7~4.0%	-1
	4.1~4.6%	-3
Operating temperature	154~157 °C	+4
	158~161 °C	+2
	162~165 °C	0
	166~169 °C	-2
	170~175 °C	-4
Operating pressure	575~624 kPa	+3
	625~674 kPa	+1
	675~685 kPa	0
	686~735 kPa	-1
	736~785 kPa	-3

Table 8.5 Cooking time ranges for coupled weight values

Weight values	Impact on cooking time	Required cooking time
+9	40~50 minutes longer	4:35~4:45 hours
+8	35~45 minutes longer	4:30~4:40 hours
+7	30~40 minutes longer	4:25~4:35 hours
+6	25-35 minutes longer	4:20~4:30 hours
+5	20~30 minutes longer	4:15~4:25 hours
+4	15~25 minutes longer	4:10~4:20 hours
+3	10~20 minutes longer	4:05~4:15 hours
+2	5~15 minutes longer	4:00~4:10 hours
+1	0~10 minutes longer	3:55~4:05 hours
0	**normal cooking time**	3:45~3:55 hours
-1	0~10 minutes shorter	3:35~3:45 hours
-2	5~15 minutes shorter	3:30~3:40 hours
-3	10~20 minutes shorter	3:25~3:35 hours
-4	15~25 minutes shorter	3:20~3:30 hours
-5	20~30 minutes shorter	3:15~3:25 hours
-6	25~35 minutes shorter	3:10~3:20 hours
-7	30~40 minutes shorter	3:05~3:15 hours
-8	35~45 minutes shorter	3:00~3:10 hours
-9	40~50 minutes shorter	2:55~3:05 hours

Table 8.5 provides the required cooking time corresponding to different weight values. If a cumulative weight value is higher or lower than those presented in the table, the system is unable to provide the

coupled cooking time. Table 8.6 shows the number of possibilities that IOSS can handle at each level. A total of 37500 different possible combinations exist at level III of the coupled system.

Table 8.6 Number of all possible situations

Individual quality inputs (level I)	Number of possible situations	Coupled quality inputs (level II)	Number of possible situations	Coupled quality inputs (level III)	Number of possible situations
Chip load	5	(wood	$(5\times4\times3$		
Chip quality	4	chips)	$=60)$		
Chip size	3			(all quality	$(60\times25\times$
Total SO$_2$	5	(cooking	$(5\times5=25)$	inputs)	$25=$
pH	5	liquor)			$37\,500)$
Temperature	5	(operating	$(5\times5=25)$		
Pressure	5	conditions)			

8.10.2 Common-sense reasoning mechanism

Generally speaking, a computer system will accept any input a user provides, even wrong input data. In a real industrial application, this is very dangerous. The data to be entered into the IOSS which describes the process status must be physically significant. The IOSS works in domains where conditions are rarely certain. The wrong input information may result in an error-prone solution.

In order to prevent a user from entering wrong data to the IOSS, a simple common-sense reasoning mechanism is built in the IOSS. For example, if the user enters the operating temperature at a value of 200° C, the IOSS will not accept this value because it is physically impossible. When this happens, the IOSS automatically requests another value. If a new value is not entered, the corresponding input variable will take the default value which is the desired value. The rules which perform the screening of non-physical numerical values are executed at the interface level so that these non-physical wrong data will not be used in the knowledge base, thus making the system more efficient and reliable.

3.10.3 Variable rules

Level I of the knowledge base is so designed that only two rules are utilized. This new technique is called variable rule, which allows the IOSS to substitute many different facts into the same general format. However, it must have a look-up table. Such a technique is used to better organize the knowledge base as well as to improve knowledge processing and to reduce running time.

Using a simple example below, we describe the function of variable rule. If all other process variables are under normal conditions, with three different values, we can represent the operating temperature variable with three production rules as follows:

(Rule 1
    IF   operating temperature is < 162 °C,
    THEN   process requires longer cooking time,
               and pulp product has good brightness.)
(Rule 2
    IF   operating temperature is > 165 °C,
    THEN   process requires shorter cooking time,
               and pulp product has poor brightness.)
(Rule 3
    IF   operating temperature is between 162 °C and 165 °C,
    THEN   process requires normal cooking time,
               And pulp product has good brightness.)

However, we can use a variable rule to represent the rules above.

(Variable rule
    IF   operating temperature is X °C,
    THEN   process requires Y cooking time,
               and pulp product has Z brightness.)

Looking up table

	X	Y	Z
1	<162 °C	longer	good
2	>165 °C	shorter	poor
3	162-165 °C	normal	good

The variable rule and look-up table represent the following knowledge:

Process requires (longer) cooking time and pulp product has (good) brightness at operating temperature (<162 °C),

Process requires (shorter) cooking time and pulp product has (poor) brightness at operating temperature (>165 °C), and

Process requires (normal) cooking time and pulp product has (good) brightness at operating temperature (>162 °C and <165 °C)).

As the knowledge base is built with arrays and matrices, the 'variable rule' system is used to find the location (loc.) in a matrix that satisfies a certain value of an input variable. The syntax of variable rules is shown in Example 3. The first variable rule structure is defined as 'rule_base_setA', and its structure is shown in line 6 of Example 3. This structure 'rule_base_setA' is used to process input variables whose value are numerical values.

Example 3: Syntax of the variable rules

```
int rule_base_setA(float xx, float range1[][2])
{
    int i;
    for (i=0; i<=9; i++)
    {
(line 6)   if (xx >= range1[i][0] && xx <= range1[i][1])
            break;
    }
```

```
    return i;
}

int rule_base_setB(float xx, float par[])
{
    int i;
    for (i=0; i<=9; i++)
    {
(line 17)  if (xx == par[i])
           break;
    }
    return i;
}
```

Another variable rule, which is defined as 'rule_base_setB', is used to process the remaining input variables, which are chip quality and chip size. Their values are defined with *words* rather than numerical values. The structure of this variable rule is shown in line 17 of Example 3. The variable rule system is coded as a separate module called 'RULEBASE.C'.

8.11 Field testing

During the development of the IOSS, the Intelligence Engineering Laboratory has organized numerous discussion and evaluation meetings for the project. Many important suggestions and much advice has helped us improve the IOSS.

After the IOSS had been developed and evaluated in our laboratory, we sent the program back to Fraser mill for field testing. Their technical control superintendent Mr Gilles Volpe wrote to us to state that the IOSS is a '*well developed expert system, and applicable into the operation for the operators and the process engineers. A lot of useful information are contained in this program with the possibility of expanding for the future Fraser's cooking model*'.

Mr Andrew M. Slowik, the manager of process systems at Fraser, had also evaluated the program. After Mr Slowik had used the IOSS, a meeting and demonstration was organized with other staff members for further evaluation. Included at the meeting were the superintendent, technical assistants, and field engineers from the digester's operation department. Attending were also representatives from central technical, plant engineering, and maintenance departments. According to Mr Slowik, the response to the IOSS was extremely encouraging. All department representatives at the demonstration had positive comments on the ease of use of the IOSS and quality of the information gleaned from its use. Many of them thought that the IOSS could be an excellent training tool for all trades and disciplines involved in digester operation, especially production personnel. Also mentioned in Mr Slowik's letter, was that several people suggested that the system could be considered as a reference for engineering/maintenance due to the completeness of information contained within the IOSS (Slowik, 1991).

We also presented the IOSS to several pulp and paper industrial companies as well as research institutes, and held discussions with application domain experts and AI development scientists to evaluate and modify the IOSS. Based on our demonstration, the industrial staff felt that such methodology is a very good model for real industrial process applications.

8.12 Illustrations

Let us select a particular batch cycle where the input variables are assigned the following data:

Chip load	32% gauge
Chip quality	good quality
Chip size	normal size
Free % SO_2	2.79%
Combined % SO_2	2.60%
Total % SO_2	5.39%
pH	3.4
Operating temperature	160°C
Operating pressure	680kPa

These data are entered into the IOSS by selecting Inputs sub-menu from Main menu #1. As an example, Figure 8.9 explains how the value for total SO_2 of the initial cooking acid is entered into the IOSS.

Once all the required data are entered, the user may first select Situation to obtain qualitative information about the process operation behavior. For instance, Figure 8.10 displays the qualitative reasoning results for operation temperature.

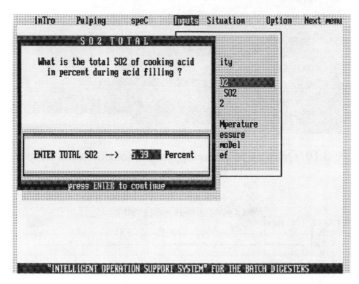

Figure 8.9 Input data for SO_2.

Since most users often prefer the graphic presentation of results, a bar-graph display system is built into the IOSS to provide the estimated cooking time. Figure 8.11 shows the estimated cooking time for each of the input variables (the coupled system at level I), where the normal range (i.e. between 3:45 and 3:55 hours) is given by a horizontal bar. The input variable whose bar is within the normal cooking time range tells us that the pulp quality is within the desired value range. Figure 8.12 displays the estimated cooking time based on the selected coupled input variables (level II), while Figure 8.13 gives the estimated cooking time based on all input variables coupled together at level III.

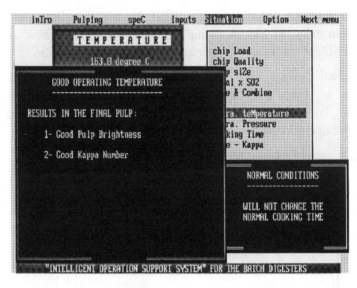

Figure 8.10 Qualitative reasoning results for operation temperature.

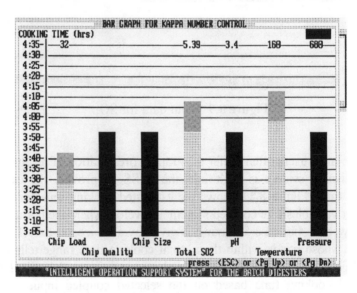

Figure 8.11 Estimated cooking time for individual input variables (I).

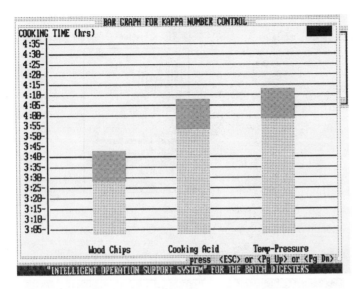

Figure 8.12 Estimated cooking time based on selected input variables (II).

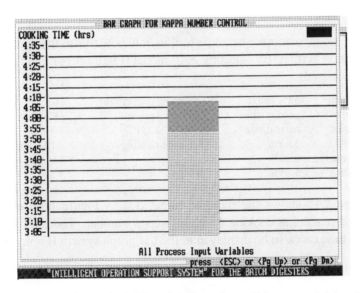

Figure 8.13 Estimated cooking time based on all input variables (III).

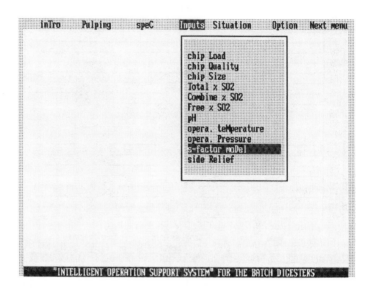

Figure 8.14 Selection to access the quantitative computation system.

Now, we give another example to illustrate how to run the quantitative computation system. By selecting S_{factor} model (Figure 8.14), the following data are entered:

Chip load	38% gauge	Operating temperature	165 °C
Chip quality	good quality	Operating pressure	680 kPa
Chip size	normal size	Total % SO_2(10)	1.89%
Free SO_2	3.05%	Time of sampling (t_1)	2:10 hours
Combined SO_2	2.90%	$S_{m1\text{-}factor}$	125 atm-hour
Total SO_2	5.95%	pH	3.5

Figure 8.15 shows the required target $S_{m\text{-}factor}$. Figure 8.16 gives the estimated remaining cooking time and kappa number. The total required cooking time can also be displayed in the bar graph system (Figure 8.17).

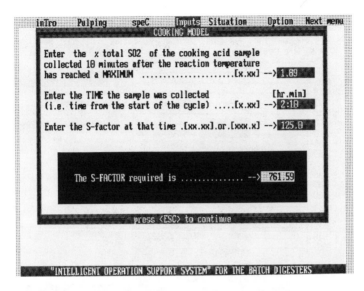

Figure 8.15 Display of the required target $S_{m-factor}$.

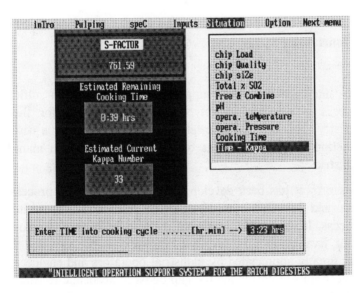

Figure 8.16 Estimated remaining cooking time and kappa number.

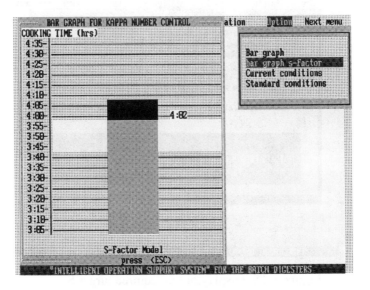

Figure 8.17 Graphical presentation of the calculated total cooking time.

8.13 Conclusions

- This intelligent operation support system for a batch sulfite pulping process is the first prototype system developed for the pulping industry. This original research will pave the way for a new research direction for future process operation and control in manufacturing industries.

- This project has been developed for a real industrial process. All the data and knowledge were collected from Fraser's sulfite pulping process. It was so designed that it could be directly employed in the plant. Evaluation at Fraser's mill was very satisfactory, and the project was greatly appreciated by industrial operators and engineers. Such a project aims at narrowing the gap between academic research and real industrial application.

- The objective of this project is to provide assistance to human operators to achieve better operation of the batch chemical pulping process. This system could be used by different users. For example, the system can be used to provide operational support for operators, to help engineers better understand the process, as well as to train new employees. It has a multimedia interface runs under a user-friendly environment. The IOSS intends to increase interaction between human operators and the pulping process by using both qualitative reasoning and quantitative computation.

- The qualitative reasoning function helps operators to select near-optimal operation conditions based on the initial process variables and operation conditions. It provides cooking time for the production of good quality pulp, supervises the process operation, and diagnoses faults.

- Based on quantitative computation, the IOSS can provide on-line supervisory control under the special process operation situations. It can alert the operator about the remaining cooking time before the drain down operation for a cooking cycle. It also estimates kappa numbers during the cooking cycle. Its results are precise and superior to those from the qualitative reasoning system. Therefore, it works in the very limited environment where chip size and chip quality are at normal conditions.

- The development of an IOSS for a real industrial application requires knowledge from different disciplines. This project utilizes knowledge from chemical engineering, computer science, artificial intelligence, and control systems.

- During the construction of the IOSS, many advanced AI techniques were investigated to improve system efficiency. They are listed below:

(a) The integrated distributed intelligence system concept is implemented in the IOSS. Qualitative reasoning, quantitative computation, as well as computer graphics are integrated.

(b) The IOSS utilizes variable rules to organize its knowledge base. Such a technique greatly reduces both memory space and running time.

(c) The IOSS has a common-sense reasoning mechanism to prevent the system from processing impossible physical data. This feature is implemented at the user interface level and prevents the users from entering wrong input data.

(d) A multimedia interface is built into the IOSS. The interface provides many features such as graphic representation of both process flow chart and reasoning-computation results, window and menu-driven environment, natural language interface, ignoring wrong input data, multi-task environment. It is implemented using C™ language, running on PC/MSDOS™ that offers the user the lowest development cost, and the higher flexibility to modify the system.

(1988) Guidelines for safe storage and handling of high toxic hazard materials. *Preventing Major Chemical and Related Process Accidents. IChemE Symposium Series No. 110,* Institute of Chemical Engineers, Rugby, UK.

Davis, R. (1987) Knowledge-based systems: the view in 1986, in *AI in the 1980s and Beyond* (eds. Grimson and Patil), MIT Press, pp.16-19.

Davis R. and Lenat, D. (1982) *Knowledge-based Systems in Artificial Intelligence,* McGraw-Hill, New York.

Davis, J. and Oliff, M.D. (1988) Requirements for manufacturing planning islands using knowledge based technology, in *Proc. of the Second Int. Conf. on Expert Systems and Intelligent Manufacturing,* (ed. Oliff, M.D.), North-Holland, New York, pp.25-42.

Dedourek, J.M., Sorenson, P.G. and Tremblay, J.P. (1989) Meta-system for information processing specification environment. *INFOR,* **27(3),** 311-37.

Denavit, J. and Hartenberg, R.S. (1988) A kinematic notion for lower-pair mechanisms based on Matrices. *ASME J. Appl. Mechanics,* 215-21.

Derby, S. (1982) Computer graphics simulation of robot arms. *Proc. CAD/CAM Technology for Manufacturing Engineering,* Cambridge, MA, pp.215-21.

Dixon, J.R. (1986) Artificial intelligence and design: a mechanical engineering: view. *Proc. of 5th National Conf. on AI,* Los Altos, pp.872-77.

Dixon, J.R. and Simmons, M.K. (1983) Computer that design: expert systems for mechanical engineers. *Computers in Mechanical. Engineering,* **2(3),** 10-18.

Dixon, J.R. and Simmons M.K. (1984) Expert system for design: standard V-belt driven design as an example of the design-evaluate-redesign architecture. *Proc. Int. Computers in Engineering Conf.,* Las Vegas, Nevada, pp.332-37.

Donat, J.S., Bhat, N. and McAvoy, T. (1990) Optimizing neural net based predictive control. *Proc. American Control Conf.,* pp.2466-71.

Press, Tianjin.

Hinde, C., West, A. and Williams, D. (1992) The use of object orientation for the classification and design of manufacturing process control systems. *Proc. of International Conference on Object-Oriented Manufacturing Systems,* Calgary.

Howie, P. (1983) Graphic simulation for off-line robot programming, *Robotics Today,* **6(1)**, 63-66.

Huang, Y.W. and Fan, L.T. (1988) Fuzzy logic rule-based system for separation sequence synthesis: an object-oriented approach. *Computers in Chem. Eng.,* **12(6)**, 601-7.

Jacobstein, N., Kitzmiller, C.T. and kowalik, J.S. (1988) Integrating symbolic and numeric methods in knowledge-based systems: current status, future prospects, driving events, *in Coupling Symbolic and Numerical Computing In Expert Systems,* II (eds Kowalik, J.S. and Kitzmiller, C.T.), New York, NY: Elsevier Science Publishers B.V., pp.3-11.

James, J.R., Frederick, D.K., Bonissone, P.P. and Taylor, J.H. (1985) A retrospective view of CACE-III: considerations in coordinating symbolic and numeric computation in a rule-based expert system. *Proc. IEEE 2nd Conf. AI Applic.,* Miami, FL, pp.532-38.

Kapp, D (1989) An analysis of distinction between deep and shallow expert systems. *Int. Journal of Expert Systems,* **2(1)**, 1-34.

Kim, J., Funk, K.H. and Fichter, E.F. (1988) Towards an expert system for FMS scheduling: a knowledge acquisition environment, in *Proc. of the Second Int. Conf. on Expert Systems and Intelligent Manufacturing* (ed. Oliff, M.D.), North-Holland, New York, pp.215-34.

Kitzmiller, C.T. and Kowalik, J.S. (1987) Coupling symbolic and numeric computing in knowledge-based systems. *AI Magazine,* Summer, 85-90.

Kletz, T. A. (1988) *What Went Wrong? Case Histories of Process Plant Disasters* (2nd edn.), Gulf Publishing Company, Houston, TX.

Kovacs, W. (1985) Previewing robotic motion with computer graphics. *Robotics Age,* **8**, 16-19.

Lamont, G. and Schiller, M.W. (1987) The role of artificial intelligence in computer-aided design of control systems. *Proc. IEEE 26th Conf. on Decision and Control*, Los Angles, CA, pp.1960-65.

Levin, H.P. (1964) Use of graphs to decide the optimum layout design of building. *Architecture Journal*, 809-17.

Li, D., Wang, Q. and Zhou, J. (1990) An engineering layout design implement system (LDS-1) based on interactive 3D graphics display technology. *Proc. of International Conference on Engineering Design ICED'90*, Dubrovnik.

Lirov, Y., Rodin, E.Y., McElhaney, B.G. and Wilbur, L.W. (1988) Artificial intelligence modeling of control systems. *Simulation,* **50**, 12-24.

Liu, H.L., Wang, Q. and Zhou, J. (1989) An intelligent CAD system for graphic generation and automatic dimensional decision of shafts. *Proc. the International Conference on Expert Systems in Engineering Applications*, Wuhan, China, pp.423-25.

Loresen, W. (1986a) An object-oriented graphics animation system, General Electric Corporate Research & Development, Technical Information Series report.

Loresen, W. (1986b) *Objective-Oriented Design,* CRD Software Engineering Guidelines, General Electric Co.

Love, T. (1985) *Message Object Programming: Experiences with Commercial Systems*, Productivity Products International, Sandy Hook, CT.

Lu, S.C.Y. (1989) Knowledge processing for engineering automation. *Proc. 15th Conf. on Production Research and Technology*, Berkeley, CA, pp.455-68.

Mann, R.W. (1965) *Computer-Aided Design*, Mcgraw-Hill Year Book Science and Technology.

Markov, L.A. (1984) Optimization techniques for two dimensional placement. *21th D.A.C.*, 652-54.

Maybury, M.T. (1992) Intelligent multimedia interfaces. *IEEE Expert*, February, 75-77.

McDermott, J. (1980) R1: A rule-based configurer of computer science. Dept. of Computer Science, Carnegie-Mellon University.

Miller, R.K. and Walker, T.C. (1988) *Artificial Intelligence Applications in Engineering,* SEAI Technical Publications, Georgia.

Moore, R.L., Hawkinson, L.B., Knickerbrocker, C.G. and Churchman, L.M. (1984) A real-time expert system for process control. *Proc. IEEE 1st AI Applic. Conf.,* Denver, CO, pp.569-76.

Mostow, J. (1985) Toward better models of the design process. *AI Magazine,* **6 (5),** 44-57.

Novak, B. (1984) Robotic simulation facilitates assembly link design. *Simulation,* **43,** 298-99.

Okuno, H.G. and Gupta, A. (1988) Parallel execution of OPS5 in QLISP. *Proc. IEEE AI Application Conf.,* pp.268-73.

Orelup, M.F. and Cohen, P.R. (1988) Dominic II: meta-level control in iterative redesign. *Proc. American Control Conf.,* Atlanta, pp.25-30.

Pang, G.K.H and MacFarlane, A.G.J. (1987) *An Expert System Approach to Computer-Aided Design of Multivariable Systems,* Springer-Verlag, New York.

Pang, G.K.H, Vidyasagar, M. and Heunis, A.J. (1990) Development of a new generation of interactive CACSD environments. *IEEE Control Systems Magazine,* **10(5),** 40-44.

Parker, H.V. (1988) Purchased chip quality control provides improved pulp quality and yield. *Tappi Journal,* December.

Parrello, B.C., Kabat W.C. and Wos, L. (1986) Job-shop scheduling using automated reasoning: a case study of the car-sequencing problem. *Journal of Automated Reasoning,* **2(1),** 1-42.

Pearson, J. and Brazendale, J. (1988) Computer control chemical plant - design and assessment framework. *Preventing Major Chemical and Related Process Accidents. IChemE Symposium Series No. 110,* Institute of Chemical Engineers, Rugby, UK.

Peng, S.W., Smith, S.F. and Howie, R. (1988) Requirements for manufacturing planning islands using knowledge based technology, *Proc. 2nd Int. Conf. on Expert Systems and Intelligent Manufacturing* (ed. Oliff, M.D.), North-Holland, New York, pp.25-42.

Prasad, S.S. (1986) Computer-aided design of optimal layouts, in *CAD/CAM/CAE for Industrial Progress,* CIFIP, North-Holland, pp.242-50.

Pressman, R.S. (1987) *Software Engineering: a Practitioner's Approach,* second edn, McGraw-Hill, London.

Qian, D. and Lu, Y. (1987) Applications of fuzzy set theory to knowledge representation and reasoning in dynamic systems. *Proc. IEEE Int. Conf. on System, Man, and Cybernetics,* Alexandria, VA, pp.1032-35.

Rao, M. (1991) *Integrated System for Intelligent Control,* Springer-Verlag, Berlin.

Rao, M. (1992) Frontiers and challenges of intelligent process control. *Engineering Applications of AI,* **5**.

Rao, M., Cha, J.Z. and Zhou, J. (1991a) PHIIS: parallel hierarchical integrated intelligent system for engineering design automation. *Engineering Applications of Artificial Intelligence,* **4(2)**, 145-50.

Rao, M., Jiang, T.S. and Tsai, J.P. (1988a) IDSCA: an intelligent direction selector for the controller's action in multiloop control systems. *Int. J. Intelligent Systems,* **3**, 361-79.

Rao, M., Jiang, T.S. and Tsai, J.P.(1988b) Integrated architecture for intelligent control. *Third IEEE International Symposium on Intelligent Control,* Arlington, Virginia, pp.81-85.

Rao, M., Jiang, T.S. and Tsai, J.P. (1989) Combining symbolic and numerical processing for real-time intelligent control. *Eng. Applic. of Artificial Intelligence,* **2**, 19-27.

Rao, M., Qiu, H., Dong, R., Kim, H., Ying, Y. and Zhou, H. (1992a) PC-oriented software for process control education. *Int. J. of Applied Engineering,* **8**.

Rao, M., Tsai, J.P. and Jiang, T.S. (1987a) Adaptive feedback testing system: case studies and applications. *Mid-West Symp. Artificial*

Intelligence & Cognitive Science, Chicago, IL, pp.43-50.

Rao, M., Tsai, J.P. and Jiang, T.S. (1987b) A framework of integrated intelligent systems. *Proc. IEEE Int. Conf. on Systems, Man, and Cybernetics,* Alexandria, Virginia, pp.1133-37.

Rao, M., Tsai, J.P. and Jiang, T.S. (1988c) An intelligent decisionmaker for optimal control. *Applied Artificial Intelligence,* **2,** 285-305.

Rao, M., Wang, Q., Coward, J. and Lamb, D.K. (1992b) Maintenance support system for truck condition monitoring. *Proc of CSchE,* Toronto. pp.242-43.

Rao, M., Ying, Y. and Corbin, J. (1991b) Intelligence engineering approach to pulp and paper process control. *77th CPPA Annual Conf.,* Montreal, Quebec, pp.195-200.

Rao, M., Zheng, X. and Jiang, T.S. (1987c) Graphic simulation: beyond numerical computation and symbolic reasoning. *Proc. IEEE Int. Conf. Systems, Man, and Cybernetics,* Beijing, China, pp.523-24.

Rembold, U., Blume, C. and Dillmann, R. (1985) *Computer Integrated Manufacturing Technology and Systems,* Marcel Dekker, New York.

Rickel, J. (1988) Issues in the design of scheduling systems. *Proc. 2nd Int. Conf. on Expert Systems and Intelligent Manufacturing* (ed. Oliff, M.D.) North-Holland, New York, pp.70-89.

Rit, J.F. (1986) Propagating temporal constraints for scheduling. *Proc. AAAI-86,* Philadelphia, pp.383-88.

Shannon, R. (1986) Intelligent simulation environment. *Proc. Intelligent Simulation Environment,* San Diego, CA, pp.150-56.

Shannon, R., Mayer, R. and Adelsberger, H. (1985) Expert systems and simulation. *Simulation,* **44(6),** 275-84.

Slowick, A.M. (1991) Letter to Mr Jean Corbin: An expert system for the operation of the batch digesters, September 18.

Smith, D. and Hubert, B. (1983) *Digesters' Operation Manual,* Fraser Inc., New Brunswick, Canada.

Smith, J. and Joshi, S. (1992) Object-oriented development of shop floor control systems. *Proc. International Conference on Object-Oriented Manufacturing Systems*, Calgary.

Smith, S.F. (1987) A constraint-based framework for reactive management of factory schedules, *Proc. of the First Int. Conf. on Expert Systems and Intelligent Manufacturing* (ed. Oliff, M.D.), North-Holland, New York, pp.113-30.

Soroka, B.I. (1984) Expert systems and robotics. *Proc. of the Robots 8th Conf.*, pp.19132-40.

Sriram, D., Maher, M.L., Bielak, J. and Fenves, S.J. (1982) Expert systems for civil engineering-a survey. Technical Report R-82-137, Dept. of Civil Engineering, Carnegie-Mellon University.

Stauffer, R.N. (1984) Robot system simulation. *Robotics Today*, **6(3)**, 81-90

Steffen, M.S., and Greene, T.J. (1986) An application of hierarchical planning and constraint-directed search to scheduling parallel processors. *Proc. IEEE Conf. on Robotics and Automation*, San Francisco, pp.1234-43.

Stephanopoulos, G. and Davis, J. (1990) Artificial intelligence in process systems engineering. *CAChE Monograph Series*, **1**.

Stefik, M. (1981) Planning with constraints (Molgen: Part 1). *Artificial Intelligence*, **16**, 111-40.

Surko, P. (1989) Tips for knowledge acquisition. *PC AI*, May/June, 14-18.

Taylor, J.H. and Frederick, D.K. (1985) An expert system architecture for computer-aided control engineering. *IEEE Proceedings*, **72**, 1795-805.

Vemuri, V. (1988) *Artificial Neural Networks: Theoretical Concepts*, IEEE Computer Society Press, Washington DC.

Volpe, G. (1991) Letter to Dr M. Rao: Digester control-exprt system. Fraser Inc., Septemper 12.

Wang, Q., Zhou, J. and Yu, J. (1989a) A method of product conceptual design using AI technology. *Proc. 5th Int. Conf. CAPE*, Edinburgh,UK, pp.200-10.

Wang, Q, Zhou, J. and Yu, J. (1989b) Decision making system of mechanical product general scheme design. *Proc. ICED'89*, pp.705-14.

Wang, Q., Li, D.S., Zhou, J. and Yu, J. (1990a) Intelligent CAD for engineering layout design. *Proc. International Conference on Engineering Design ICED'90*, Dubrovnik.

Wang, Q., Yang, H., Zhou, J. and Yu, J. (1990b) QUINT: A problem solving strategy for mechanical system concept design, Part I and Part II. *Proc. Advance in Design Automation*, Chicago, IL, pp.331-43.

Watanbe, K., Matsuura, I., Abe, M., Kubota, M., and Himmeblau D. (1989) Incipient fault diagnosis of chemical processes via artificial neural networks. *AIChE Journal*, **35(11)**, 1903-12.

White, R.B., Read, R.K., Read, M.W., Koch, M.W. and Schilling, R.J. (1989) A graphics simulator for a robotic arm. *IEEE Trans. Education*, **32**, 417-29.

Wiener, R. and Sincovec, R. (1984) *Software Engineering with Modula-2 and Ada*, Wiley & Sons, New York.

Yan, M. (1988) Computer-aided space layout design. *Microcomputer System*, **9(12)**, 1-14.

Winston, P. (1975) Learning structural descriptions from examples, in *The Psychology of Computer Vision*, McGraw-Hill, New York, pp.157-209.

Wright, M.L., Green, M.W., Fiegl, G. and Cross, P.F. (1986) An expert system for real-time control. *IEEE Software*, March, 16-24.

Wong, F.S., Dong, W. and Blanks, M. (1988) Coupling of symbolic and numerical computations on a microcomputer, *AI in Engineering*, **3**, 32-38.

Yeomans, R.W. (1984) *CIM and the ESPRIT Programs*.

Yeomans, R.W., Chaudry, A. and Tenhagen, P. (1985) *Design Rules for A CIM System*, North-Holland.

Yong, Y.F. Gleave, J.A., Green, L. and Bonney, M.C. (1985) Off-line programming of robots, in *Industrial Handbook on Robotics*, Wiley & Sons, New York.

Yorston, F.H. and Liebergott, N. (1965) Correlation of the rate of sulphite pulping with temperature and pressure. *Pulp & Paper Canada,* May, 272-78.

Zadeh, L.A. (1978) Fuzzy sets as a basis for a theory of possibility. *Fuzzy Sets and Systems,* **1**, 3-28.

Zeigler, B.P. (1984) Multifaceted modeling methodology: grappling with the irreducible complexity of systems. *Behavioral Science,* **29**, 169-178.

Zhou, J., Duan, J. and Wang, Q. (1989a) A study of evaluation subsystem for scheme design expert system. *Journal of HUST,* **17(2)**, 10-24.

Zhou, J., Wang, Q. and Yu, J. (1989b) Mechanical product general scheme optimization design and intelligent CAD. *Journal of Mechanical Engineering,* **3**, 56-61.

Further reading

Akrif, O. and Blankenship, G.L. (1987) Computer algebra for analysis and design of nonlinear control systems. *Proc. of American Control Conference*, Atlanta, GA, pp.547-54.

Butler, C.W., Hodil, E.D. and Richardson, G.L. (1988) Building knowledge-based systems with procedural languages. *IEEE Expert*, Summer, 47-59.

Colgren, R.D. (1988) Development of a workstation for the integrated design and simulation of flight control systems. *Proc. NAECON 88*, Dayton, Ohio, pp.380-87.

Cox, B. (1985) Software ICs and objective-C. *Unix World,* Spring.

Cox, B. (1986) *Object Oriented Programming,* Addison-Wesley.

Cruise, A., Ennis, R., Finkel, A., Hellerstein, J., et al. (1987) YES/L1: integrating rule-based, procedural, and real-time programming for industrial applications. *Proc. IEEE Third AI Application Conf.*, pp.134-39.

Dong, R., Rao, M., Bacon, H. and Mahalec, V. (1991) An expert system for safe plant startup. *PSE'91*, IV.4, pp.1-4.15.

Doyle, J. (1979) A truth maintenance system. *Artificial Intelligence*, **12(3)**, 231-72.

Files, R. and Kehler, T. (1985) The role of frame-based representation in reasoning. *ACM*, **28**, 904-20.

Goldberg, A. and Robson, D. (1983) *Smalltalk-80: The Language and Its Implementation*, Addison-Wesley, Reading MA.

Gundersen, T. (1991) Achievements and Future challenges in industrial design application. *Proc. of PSE'91*, Quebec, pp.I.I.1-I.I.31.

James, J.R. (1987) A survey of knowledge-based systems for computer-aided control system design. *Proc. of American Control Conference*, Minneapolis, MN, pp.2156-61.

Jones, M. (1985) Applications of artificial intelligence within education. *Comp. & Math. with Appls.*, **11(5)**, 517-26.

Kidd, P.T. (1990) Organization, people and technology: advanced manufacturing the 1990s. *Computer-Aided Engineering*, **7(9)**, 149-52.

Lamit, L.G. and Paige, V. (1987) *Computer-Aided Design and Drafting (CADD)*, Merrill Publishing Company, London.

Lizza, C. and Friedlander, C. (1988) The pilot's associate: a forum for the integration of knowledge based systems and avionics. *Proc. NAECON 88*, Dayton, OH, pp.1252-58.

Manoocheiti, S. and Seireg, A. (1987) Computer-aided generation of an optimum machine topology for specified tasks. *CIME*, **6(3)**, 10-24.

Marchand, D.A. (1988) Strategic information management: challenges and issues in the CIM environment, *Proc. 2nd Second Intern. Conf. on Expert Systems and Intelligent Manufacturing* (ed. Oliff, M.D.), North-Holland, New York, pp.11-24.

Moses, J. (1967) Symbolic Integration, Ph.D. Thesis, Massachusetts

Institute of Technology, Cambridge, MA.

Rao, M. and Jiang, T.S. (1988) Simple criterion for testing non-minimum-phase systems. *Int. J. Control,* **47**, 653-56.

Rao, M. and Jiang, T.S. (1989a) Expert-aided process control laboratory. *Intern. J. Applied Eng. Education,* **6(2)**, 227-31.

Rao, M. and Jiang, T.S. (1989b) Expert systems for process control: a survey. *Proc. Instrument Society of American Southeastern Conf. '89,* Pensacola, FL.

Rao, M., Cha, J.Z. and Zhou, J. (1990) New software platform for intelligent manufacturing. *Proc. AAAI 90 Workshop on Intelligent Manufacturing Architecture,* Boston, MA, pp.62-65.

Rao, M., Theisen, C. and Luxhoj, J. (1990) Intelligent system for air-traffic control, *5th IEEE Intern. Symp. on Intelligent Control,* Philadelphia, PA, pp.859-63.

Rao, M., Jiang, T.S. and Tsai, J.P. (1990) Integrated intelligent simulation Environment. *Simulation,* **54(6)**, 291-95.

Robert, E.F. (1988) Reasoning with world and truth maintenance in a knowledge-based programming environment, JACM, **31(4)**, 382-401.

Sanchez, J. (1990) *Graphics Design and Animation on the IBM Microcomputers,* Prentice Hall, New Jersey.

Saridis, G.N. (1983) Intelligent robotic control. *IEEE Trans. Automatic Control,* **AC-28**, 547-57.

Saridis, G.N. and Valavanis, K.P. (1988) Analytical design of intelligent machines. Automatica, **24**, 123-33.

Shirley, R.S. (1987) Some lessons learned using expert systems for process control. *Proc. American Control Conf.,* pp.1342-46.

Slagle, J.R. (1963) A heuristic program that solves symbolic integration problems in freshman calculus. *JACM,* **10**, 507-20.

Su, S.Y.W. and Lam (1990) Object-oriented knowledge base management technology for improving productivity and Competitiveness in Manufacturing, *NSF Grantees Conf. on Design and Manufacturing*

Systems, Tempe, AZ.

Talukdar, S.N., Cardozo, E. and Leao, L.V. (1986) Toast: power system operator's assistant. *IEEE Computer,* July, 53-60.

Tsuto, K. and Ogawa, T. (1991) A practical example of computer integrated manufacturing in chemical industry in Japan. *Proc. PSE'91*, Quebec, II.2, pp.1-2.24.

Yu, J., Zhou, J. and Wang, Q. (1987) Mechanical product general scheme design CAD based on expert system technology. *Proc. Advance in Design Automation*, Boston, MA, pp.310-14.

Zhang, Z. and Rice, S.L. (1989) Conceptual design: perceiving the pattern. *CIME*, **11(7)**, 58-60.

Zhao, H.C. (1986) *Level Analysis Methodology*, Science Press, Beijing.

Index